I0824322

Praise for *Cheese Trekking*

"*Cheese Trekking* is an important book about food sovereignty, and it is a pleasure to read. Trevor Warmedahl's stories offer profound lessons about cheesemaking and life. His reflections flow with a certain poetry, featuring vivid, expressive descriptions of places, people, and animals and the unique flavors, textures, smells, and qualities of different cheeses and dairy concoctions. This book places cheesemaking within the broader context of place-based subsistence, where cheese, the animals whose milk it is made from, and the plants and land they graze upon are all elements of an elaborately orchestrated web of cultural and biological relationships. I love this book!"

—**Sandor Ellix Katz**, author of *The Art of Fermentation* and *Wild Fermentation*

"With *Cheese Trekking*, Trevor has penned what may be the most poetic, touching, and world-expanding treatise on fermentation and symbiosis to have ever been put to the page. This book will change not only how you think about cheese—and the humans, microbes, and animals that cocreate it—but also how you think about life. Anyone who cares about where their food comes from needs to read this book."

—**David Zilber**, chef and food scientist; coauthor of *The Noma Guide to Fermentation*

"Trevor Warmedahl's book is brimming with hard-earned knowledge about locally crafted cheeses and pastoral traditions. I've never come across a book that is part adventure travelogue, part methods manual, and part record of ancient foodways while at the same time a wise treatise on how to see our place on this Earth and how to live well. An important and timeless book."

—**Helen Whybrow**, author of *The Salt Stones*

"In this anarcho-pastoralist masterpiece, Trevor Warmedahl looks to the old shepherds and cheesemakers for hope—on how to save our cheese, but also our humanity. A wisdom-filled window into alternative dimensions of cheesemaking, this book builds reverence for the sanctity of milk and will help us reclaim our greatest food: natural cheese."

—**David Asher**, author of *Milk Into Cheese* and *The Art of Natural Cheesemaking*

"Trevor Warmedahl is part tour guide, philosopher, naturalist, cheesemaker, activist, ecologist, and food guru. I've been lucky enough to follow his extraordinary travels over the last decade. Here at last he's gathered his thoughts in a gorgeously composed book about the entangled lives of animals, humans, and landscapes. In his tale of pastoralist people the world over, he offers us sober hope and strategies for a way out of our present dystopia. Pointed, impassioned, and at times hilarious, this book has taught me so much. A must-read for anyone interested in food and culture and surviving on a burning planet."

—**Brad Kessler**, author of *Goat Song* and *North*

"*Cheese Trekking* provides a powerful new lens through which to view our world—one that will give you a new way of thinking about people, animals, and food."

—**Dan Saladino**, author of *Eating to Extinction*

"Trevor Warmedahl is the Jack Kerouac of cheese, sharing his renegade passion for the unprocessed and unpasteurized as he treks around the dairy-sphere—often staying in yurts, on couches, and under the stars—to learn about traditional cheesemaking. For years, I followed his zigzag trail via Instagram, awestruck by his observations and images, always curious to know more. Now, his book brings us fully into his journey, his poetic musings, and his ongoing questions about the future of milk, microbes, and humanity itself.

"If it's possible to write a cheese book that a reader cannot put down, Trevor Warmedahl has done it! What an inspiring, intrepid, exhilarating book."

—**Tenaya Darlington**, author of *Madame Fromage's Adventures in Cheese*

"In this book of remembering, Trevor Warmedahl explores the edges of our world, far beyond the heart of capitalism's sanitized reach, for the traces of people and practices that are the link to our collective past and to what makes us human. Trevor takes us to places where the people's contract with the wild is still intact, and where the sacred link between people, their animals, and the land measures in centuries and millennia. This book is an awakening, a journey to the heart of something, and a reminder that if you travel just a little bit further and hunger a little more deeply, you will find humanity out there on the fringes, where it has been all along. I *love this book.*"

—**Mateo Kehler**, Jasper Hill Farm

"Trevor traces the history of cheesemaking through vivid stories of remote dairies and shepherds. He shows how food reflects who we are and who we might become and inspires us to imagine a society where our connection to food is guided by health and meaning rather than price, profit, and convenience."

—**Martin Rosberg**, natural cheese educator and advisor

"*Cheese Trekking* is a remarkable book by a remarkable man who set out to find traditional cheesemaking cultures all over the world and discovered culinary treasures in the process. Nobody interested in local food should miss it."

—**Ilse Köhler-Rollefson**, author of *Hoofprints on the Land*; founder, League for Pastoral Peoples

"Trevor Warmedahl's extraordinary travelogue shows that the future of cheese is not about preserving traditions in aspic, but about drawing from them, learning how microbes, farming practices, and the environment once worked together to create distinctive flavors. *Cheese Trekking* makes a powerful case for rebuilding terroir and embracing microbial diversity to shape cheese cultures that are at once rooted and new."

—**Bronwen and Francis Percival**, authors of *Reinventing the Wheel*

"Grounded in his travels across mountains and dairies, Trevor Warmedahl's book shines a light on practices long relegated to the margins, asking us to rethink the ingredients and ways used to craft good cheese. It expands our sense of where wisdom in food comes from, reaching far beyond the usual cultural references. Like Trevor's journeys documented on social media, this book deepens, expands, and unsettles our understanding of cheese—and transforms the way we taste the world."

—**Carlos Yescas**, food advocate; social enterprise consultant

Cheese Trekking

Cheese Trekking

HOW Microbes, Landscapes, Livestock, AND Human Cultures SHAPE TERROIR

Trevor Warmedahl

Chelsea Green Publishing
White River Junction, Vermont
London, UK

First published in 2026 by Chelsea Green Publishing | PO Box 4529 | White River Junction, VT 05001 | West Wing, Somerset House, Strand | London, WC2R 1LA, UK | www.chelseagreen.com
A Division of Rizzoli International Publications, Inc. | 49 West 27th Street | New York, NY 10001 | www.rizzoliusa.com

Publisher: Charles Miers
Deputy Publisher: Matthew Derr
Project Manager: Natalie Wallace
Developmental Editor: Anna Bliss
Copy Editor: Laura Jorstad
Proofreader: Ashley Davila
Indexer: Linda Hallinger
Designer: Jenna Richardson

ISBN 978-1-64502-298-5 (hardcover) | ISBN 978-1-64502-299-2 (ebook) | ISBN 978-1-64502-300-5 (audiobook)
Library of Congress Control Number: 2025037152 (print)

Our Commitment to Green Publishing

Chelsea Green sees publishing as a tool for cultural change and ecological stewardship. We strive to align our book manufacturing practices with our editorial mission and to reduce the impact of our business enterprise in the environment. We print our books on chlorine-free recycled paper, using vegetable-based inks whenever possible. This book may cost slightly more because it was printed on paper that contains recycled fiber, and we hope you'll agree that it's worth it. *Cheese Trekking* was printed on paper supplied by Sheridan that is made of recycled materials and other controlled sources.

Authorized EU representative for product safety and compliance
Mondadori Libri S.p.A. | www.mondadori.it
via Gian Battista Vico 42 | Milan, Italy 20123

Printed in the United States of America.
10 9 8 7 6 5 4 3 2 1   26 27 28 29 30

To all the knowledge holders out there
who are doing the difficult and unacknowledged work
of keeping the seeds of their place and people alive.

To begin remembering our indigenous belonging on the earth back to life, we must metabolize as individuals the grief of recognition of our lost directions, digested into a valuable spiritual compost, that allows us to learn to stay put without outrunning our strange past. Let us get small, unarmed, brave, and beautiful.

—Martín Prechtel,
The Unlikely Peace at Cuchumaquic (audiobook edition)

Contents

— OVERVIEW OF INGREDIENTS —

Milk, Culture, Rennet, Salt

Before beginning my tales of cheese trekking, I'd like to give an overview of the ingredients used in cheesemaking. This will equip the reader with some of the general science involved and lay the foundation on which the chapters can build. I will explain the four ingredients through the lens of natural cheesemaking, with an emphasis on how these elements have gone from regionally sourced to globally traded, leaving cheesemakers divorced from the primary materials of their craft. Throughout this book, I'll use term *milk microbiota* to refer to the microbial communities indigenous to raw milk, the bacteria and fungi that are the foundation of healthy raw milk and can be cultivated to make cheeses expressive of the places they are made. Multiple **microbiota** (the term can be singular or plural) are found in various environments (barns, pastures, wood equipment) or on various body parts (skin, teat, mouth). Please note that key terms are set in bold at first mention and further explained in the glossary at the end of the book.

1. Milk

As mammals, we consume colostrum and milk as our original foods. They contain all that our newborn bodies need to start developing a functioning immune system, which is not some distinct entity but an integral aspect of having an intelligent body connected to an intelligent planet. Lactation can be thought of as a continuation of gestation after the umbilical cord has been severed. Postpartum, the infant and mother continue to maintain interconnected bodies via the infant's frequent nursing. Milk is not just nutrients; it also carries microbes that seed and feed the infant's developing digestive tract microbiota. The mother is in a sense transferring her immune system, her body, her life to her child, in the acts of gestating, birthing, and nursing. There is no discrete barrier between mother and child; milk is their connection, dissolving the illusion of separateness that is foundational to modern paradigms heralded as science.

Milk contains both pre- and probiotics. The prebiotics include sugars, mainly **lactose**, but also human milk oligosaccharides (HMOs), which have equivalents in other animals. These chains of sugars are not digestible by our bodies; they are food (prebiotics) for the beneficial bacteria (probiotics) that are in milk and in our gut microbiota. By evolutionary design, milk is a food that contains sugars and the bacteria that ferment them. Ferment is what milk is designed to do. For a long time, it was thought that milk was sterile inside the animal, and that it was only inoculated with microbes as it left the body, primarily by those that reside on the surface and orifice of teats. Recent research, part of a shifting paradigm in microbiology, shows that there are also internal pathways seeding the milk with microbes before it is secreted. These are transferred from the mother's digestive tract into her mammary glands, with some perhaps being native to that environment. These internal sources can make up roughly 20 percent of the initial milk microbiota, which rapidly begins undergoing changes as it is exposed to the outside world.

For our discussion of milk's fermentation into cheese, the teat microbiota is especially important. A community of microbes live here, feeding on the milk that washes the skin every time a baby suckles, living in various niches around and inside the orifice where milk emerges. This is an interface zone between the mother's body and the outside world, a border region where the microbial balance of raw milk shifts and interacts with a wider environment. As milk leaves the teat, the diversity of potential sources of the milk microbiota explodes, to include the milking equipment, the material that animals lie down on, their feces, the barn, and the milking and cheesemaking spaces. A certain paradigm still tends to view microbes in milk as being contaminants, and it is even upheld by many producers of raw drinking milk, who attempt to produce milk with the lowest possible bacterial counts. The idea is that things like the teat microbiota contaminate the otherwise inert milk, so they should be neutralized. As I will argue, this view is misguided and dangerous. Milk and microbes are inseparable. Microbes are not just an integral part of milk but also the basis for healthy, functioning bodies, soils, landscapes, and planets. Life without microbes is the delusion of a sociopathic, distorted worldview. *Contamination* is another word for "life." We live on a planet "contaminated" with biology, despite ongoing attempts to sanitize and pasteurize the earth.

Milk can be considered an emulsion in which fat, protein, minerals, sugars, vitamins, and microbes are diffused throughout water. These constituents come from various sources, mainly the mother's body channeling the inputs from her environment. Milk is made from the water she drinks, the plants she eats, the ground where she sleeps, the air she breathes. Her body digests food and transfers the constituents into her blood, which is filtered, and the pieces are put back together in water, becoming milk. Milk is a composite of a place, assembled in the mother's body, then passed to her child, who is nourished by this predigested place, which is wrapped up neatly in the perfect package. The child's body knows exactly what to do with milk. It turns it into cheese.

I will refer to milk in its unfermented state as **sweet milk**, meaning that its sugars are intact, their fermentation into acids just starting. It is sweet milk that many of us consume, from the grocery store or raw from the farm, but the widespread consumption of sweet milk by adults is likely a modern phenomenon, born with the advent of refrigeration. For most of human history, and in many places to this day, milk is more commonly consumed in a sour, fermented state. When sweet milk is milked into and stored in porous vessels (such as pottery, wood, or animal hides) that harbor milk-fermenting bacteria, it quickly sours and eventually coagulates due to the buildup of lactic acid. Milk in this sour state is easier for us to digest, the nutrients are more bioavailable, and the milk is less prone to infiltration by unwanted microbes. The modern obsession with sweet milk requires incredible amounts of energy to keep milk from doing what it does naturally and spontaneously—that is, ferment, turn sour. By chilling milk as fast as possible, storing it cold, then pasteurizing it and homogenizing it to prevent cream from rising, we are working against nature and creating a debacle in line with the results of other aspects of modern industrial agriculture. How many burnt bridges could be rebuilt if we instead allowed milk to flow down the well-worn canyons that life has carved, encouraging it to sour and ferment, allowing it to fulfill its purpose and become cheese?

2. Culture

The main sugar in milk is lactose, which is made up of two simpler sugars. Sweet, fresh, unpasteurized milk left in a jar at room temperature (60–70°F, 15–21°C) naturally begins to sour as lactose gets consumed by bacteria into a range of byproducts, most important among them lactic acid. The buildup of lactic acid can be detected on the tongue and in the nose, for as the milk sours, its aroma and taste change. This fermentation of lactose into **lactic acid** is performed by a category of bacteria that will be key players in every cheese we will discuss: **lactic acid bacteria (LAB).** In the situation described above, the LAB were not added but were already present in the raw milk. These bacteria have also been cultivated as what are called undefined or **heirloom cultures** or **starters** (such as yogurt), which can be used indefinitely as long as they are kept fed or frozen. I use the umbrella term **naturally fermented** to include cheeses made with heirloom starters such as yogurt, **kefir**, or **clabber**, as well as those made with **whey** (the protein-rich liquid that remains after cheese is made) and wooden cheese vats. Any cheese made without **commercial starters** is naturally fermented. I also include under this umbrella those cheeses made with no tangible starter added.

Lactic acid bacteria play a key role in many fermented foods and beverages, such as sourdough, cured meats, kombucha, and some types of beer. They are prominent in sauerkrauts and other lacto-fermented vegetables and pickles, where they are not added but are part of a range of microbes found living on fruits and vegetables, similar to the situation in milk. In the case of lacto-fermented vegetables, the LAB are selected for by salt: they can handle the 1 to 3 percent that is added, while other microbes can't. Once the LAB multiply and produce an acidic environment, the foods are preserved and made more delicious and digestible. These approaches represent a collaboration and relationship with microbes, in which we learn to keep modified microbial communities healthy by feeding and caring for them, much as we do for livestock or crops.

Most of the cheese produced globally is made by adding freeze-dried, packaged starter cultures to milk that is either raw or pasteurized. These contain single strains or blends of bacteria that have been isolated out of milk or fermented dairy foods, then propagated in an industrial laboratory setting. They have become ubiquitous in the United States, the United Kingdom, and most of Europe, where their use is rarely questioned. Even home-scale hobby cheesemakers tend to see them as necessary. Globally, they are increasingly seen as the only safe, predictable, consistent, and modern way to make cheese. These select-strain cultures are commonly referred to as commercial or **direct vat inoculation starters (DVI)**. I will use the terms *commercial starters* and *DVI* to reference them. Unlike heirloom starters, they are single-use. You cannot make a batch of yogurt from them and then keep the same strains alive indefinitely by using a bit of the yogurt to start the next batch. They eventually become inactive. This means that cheesemakers who use them must continually purchase them. Subsidiaries of two biotech companies produce almost all the world's supply, and I've seen their little packages in the most unimaginable places, far from drivable roads, in remote mountainous settlements that until recently were fairly isolated from the outside world. The tentacles of the corporate food system stretch widely, attempting to leave no refuges of food sovereignty untouched and free from their imperial grasp.

This is a new way to go about fermenting milk that wasn't introduced until the 1970s. The use of DVI starters goes hand in hand with the pasteurization of milk, which reduces the level of microbial life in milk to a minimum, creating the fantasy of a blank canvas, of control over life. Pasteurization doesn't actually kill all the microbes in milk, resulting in a sterile liquid. The stated intention is to reduce the levels of potential pathogens to an acceptable minimum. Pasteurized milk is cooled to a specific temperature and packaged; then commercial starters are added allowing the lab-grown strains to multiply in the milk, where they outcompete any surviving native milk microbes. DVI starters are also used with raw milk, often at a dosage that effectively overwhelms the native microbes, preventing them from expressing. Native raw milk microbes can still express if the dose of commercial starter is low enough and the milk hasn't had its ecology irrevocably altered by cold storage.

I don't feel that cheeses made from commercial starters are inherently inferior to those made with natural ones. There are many superb cheeses made from DVI, and uninteresting ones made naturally. That said, I am philosophically and politically opposed to commercial starters. I feel they are designed to meet the needs and address the ills of a broken food system, and that we should be free to ferment foods via alliance with microbes, with whom we live in symbiotic relationships. I am opposed to microbial imperialism, biopiracy, and the impacts of transnational corporate capitalism on our foods, bodies, and lives. Living in communion with microbes, dairy animals, landscapes, one another, and the planet is our birthright. That list starts with microbes, and that's where my resistance to the corporate food system starts.

It is important to note that just because the powdered contents of these packages are being added does not mean they are successfully inoculating the milk. Commercial cultures require storage at specific, ultra-low temperatures to maintain viability. The package is designed as a single-use inoculant; once the seal is broken, the viability and purity of the culture is questionable. It has been exposed to the real world—the scary, microbe-filled one we all live in. This means that makers who store open pouches in a freezer and then weigh out what they assume is a standard amount cannot guarantee the efficacy of this approach. I do think that they generally work in commercial settings, because my senses tell me fermentation has occurred, and I can taste the DVI cultures in the cheese. They usually taste kind of generic. Standardized and familiar. Consistently boring. But it took exploring the fringes of cheese, seeking its wildest incarnations, for me to be able to discern this. The highly impactful cheeses I cover in this book were not all delicious, but they were far from boring, and getting to know them expanded my palate and showed me some of the infinite range of milk's potential. I still eat store-bought sour cream, and it tastes pretty good. It gives me what I expect. But I know it can be so much more, because I've tasted sour cream that made me shudder and weep, that had something to say, that had **terroir**.

The thing that sets naturally fermented cheeses apart is that they utilize an active community of microbes, a **SCOBY**, which is an acronym for "symbiotic culture of bacteria and yeast." This SCOBY can be maintained as an heirloom starter, in which case the community is not static but shifts with the conditions in which it lives, responding to how humans cultivate it. In the case of no-added-starter cheeses, the SCOBY is based on the raw milk microbiota, which interacts with other microbiota of the farm and cheesemaking spaces, even the bodies of cheesemakers. Nothing is done to intentionally cultivate a SCOBY and add it to milk; it is more akin to a wild fermentation. In both cases, the microbes involved have never been removed from their evolving communities and the various environments through which milk moves as it becomes cheese. In contrast, commercial starters have been isolated and propagated in a space that is claimed to be separated from the outside world, hypersanitized. They are frail and have obviously lost a lot of the natural resilience of wild things by being domesticated, removed from the communities in which they once thrived and adapted. It is the community that gives individual microbial species strength. Without their SCOBY, standing naked in the cold, they have lost the support network and defensive capability of the tribe. If something must be cultivated in a clean, sterile box in which life has supposedly been eliminated, then stored in sealed packages, it says a great deal about its ability to stay alive in the world where we live, with all other living organisms. The world beyond the walls we construct. The world of nature, the wild, biology. The only world we have.

3. Rennet

Rennet is perhaps the most controversial and misunderstood ingredient in cheesemaking. The original form is animal rennet, which is sourced from the abomasa of ruminants, generally those still drinking their mother's milk. The **abomasum** is the fourth compartment of the ruminant digestive tract. Similar to our stomachs, it's a highly acidic environment where food is broken down and compacted before it passes into the intestines. It secretes enzymes that coagulate liquid milk, making nutrients more available for digestion by the young mammal. Milk is designed to acidify and coagulate in the abomasum, to become a solid. This process is an essential part of mammalian biology, with mothers passing nutrients from their bodies to those of their children. Liquid milk is just a temporary form taken to allow the transfer; its destiny is to become cheese. By evolutionary design, cheese is what milk becomes.

If you could see the digestive tract of a very young, milk-drinking ruminant, you would see that the abomasum at this stage is by far largest organ, while the first three compartments are tiny. The rumen, reticulum, and omasum will increase in size later, as they are involved in the digestion of plants. In the first couple weeks of life, the baby is receiving its nutrition solely from its mother's milk. Because of this, the milk bypasses the first three sections, via a shortcut that closes off the passage into the rumen, forming a straight shot to the abomasum. The dominant enzyme secreted at this stage is one that is designed to coagulate milk: chymosin. It is released by the lining of the abomasum and causes the warm milk, fresh from the mother's teat, to coagulate over the course of about an hour.

The major protein in milk is casein, which exhibits a unique property. When casein is exposed to the chymosin secreted by the abomasum, a shift takes place causing the proteins formerly diffused throughout liquid milk to aggregate. Casein micelles start linking together, forming a lattice structure.

What was an emulsion in which all the components were floating freely in water begins to shift phase, like water turning to ice. The weblike lattice formed among casein micelles reaches a critical mass, and the milk suddenly gels. The other components in the milk (fats, vitamins, minerals, and water) get trapped in the matrix of the **coagulum**: coagulated, gelatinous milk. Ribbons of muscles surrounding the abomasum contract in an action known as peristalsis, which our stomachs do as well, sometimes resulting in an audible rumbling as foods are rearranged and compacted. This squeezing is like cutting the coagulum that has formed, breaking it into smaller **curds** that can pass into the intestines, where the major extraction of nutrients takes place. I'm not using metaphors from cheesemaking here—this is cheesemaking. This coagulated milk is the original cheese. Humans didn't invent cheesemaking. It is a biological process that we learned how to replicate outside the body.

Rennet is important for the concepts of this book beyond its chemistry and impact on cheese flavor development. Discussing it leads directly to the ethical dilemmas of consuming milk and cheese, and the conversation around the killing and eating of animals of all ages that dairying involves. The young animals are not being killed simply so their abomasa can be used in cheesemaking. They are being killed so that the milk they would otherwise be consuming can be taken by humans, to make into foods we will eat. They are being sacrificed for this reason, and then their bodies become meat, their hides can be tanned, and their abomasa can be utilized to coagulate the milk that they would have consumed and made into cheese themselves. We have made a major intervention into the life cycles of ruminant biology, changed their evolutionary paths, and taken on responsibility for the lives of these animals. In my view, this can be done ethically, starting with the consumption of the sacrificial animal and the utilization of all the gifts it offers us. Dairying can be done in a way that honors life and acknowledges respectfully the role of humans in the death of the animals we care for. The modern industrial way of raising livestock, obtaining milk, and making cheese has become the opposite of this, a sacrilege that is difficult to fully take in, so vast is its destructive folly and profaning of life.

There are places in the world where dairy livestock are not sacrificed at a young age but allowed to grow to maturity, drinking their mother's milk before transitioning to a plant diet. The young are allowed to suckle, while humans intervene to milk the mothers as well, sometimes referred to as milk sharing. Even in this situation, eventually most of these young animals will be killed and consumed by humans. Even the females who are raised as dairy animals, giving birth and being milked themselves, will eventually reach the end of their lives and become food as the circle of life rolls on, never stopping. Death and killing are inherent to dairying—and, I would argue, to life. It seems indicative of a huge disconnect and ethical blind spot to talk about vegetarian milk or cheese, as if somehow by not using animal-based rennet, no death is involved in the making of cheese. I feel it would be healthier to look directly at the reality of this matter, to talk about ways we can return dignity and respect to how we care for dairy animals, how we come to drink their milk and eat their flesh. We can do so in a manner that honors the webs of life that include us all.

The industrial manner of raising livestock for meat and milk is a topic that many of us are rightfully concerned about. This appalling approach represents a complete absence of morality and avoidance of responsibility for the incredible sacrifice we are asking livestock to make. Like many vegans and vegetarians, I am reacting against this system and its inhumane and environmentally destructive foundations. I do so not by saying we should stop consuming animal products, but by asking how we can raise dairy livestock in a manner that is humane and ecologically beneficial. My work is in part looking for the places where older ways of living with livestock continue to accomplish this. At the same time, I'm exploring new experiments in ways of raising and milking livestock while rebuilding fertile soil and functioning ecosystems through human-livestock-microbe alliances. For me, the conversation about milk, cheese, and responsible grazing must speak to the killing and eating of animals. Milk and meat are inseparable. There's no cheese, lifestyle, or diet that is free of death.

I divide rennet into three broad categories: animal, synthetic, and plant. Animal rennet is the original; the enzymes it contains evolved to coagulate milk and do so very efficiently. There are many cheeses that are sold as vegetarian, which usually means they use rennet that is not of animal origin. This rennet is usually not of plant origin, either, but is a product of the same paradigm that gives us commercial starter cultures, made by the same corporate entities.

About 90 percent of the cheese made in North America is derived from products often referred to as vegetarian rennet, and the rest of the world is following suit. A more accurate term is 100 percent chymosin, or **fermentation-produced chymosin (FPC)**. An even better descriptor would be genetically modified rennet, because these products result from splicing genes from a calf's abomasum onto a fungus, creating a genetically modified organism that secretes pure chymosin. This was the first bioengineered food additive approved by the FDA, entering the food system of the United States in 1990. FPC has come to dominate cheesemaking because it is cheap, accessible, and of standardized strength. It's the white sugar of cheesemaking. Its widespread adoption has been treated as a victory by animal rights activists who targeted the American veal and rennet industries in the 1970s and '80s. The thought was that, with this product, baby calves would no longer need to be killed to make our cheese. To this day, many vegetarians choose to not eat cheese made from animal rennet, while consuming those made from FPC, with the belief that this makes their cheese death-free. Young male calves are still being killed, often at birth, if not sold at auctions to be raised for meat. The death remains and, to my mind, the ethical implications are even more alarming for being essentially hidden.

True plant rennets are extracts that contain enzymes that can coagulate milk. These enzymes didn't evolve to coagulate milk, and the coagulation they induce is not equivalent to that of chymosin. They tend to form a weaker curd that is more fragile and yields less cheese. Plant rennets can lead to bitterness in cheese, caused by the cascade of flavor development starting from the introduction of botanical enzymes. Some of these flavors can be quite pleasant, and there are ways of working around problematic bitterness. Plant rennets can also lead to a breakdown of the interior of the cheese (the paste) that is hard to control or predict. Textural changes are accelerated, with cheeses softening and even liquefying under the influence of these highly reactive enzymes. This volatile ripening can be desirable, and makers can learn to steer it, slow it down, make it a part of their process, but plant rennets are tricky and work best with accumulated experience.

By far the most common plant rennet is thistle, which is often used in sheep's milk cheeses on the Iberian Peninsula. Portugal is the epicenter of this, home to many famous cheeses made with thistle rennet. More accurately, thistle rennet is typically made with the flowers of cardoon thistle. This plant resembles an artichoke but has been bred to produce large, edible stems that are delicious when braised, having a pleasantly bitter earthiness. The purple stamens from flowering cardoons can be soaked in water to make a rennet, which I cover in chapter 13.

There is a long list of plants that can coagulate milk, and a shorter one of those with established history of cultural usage as rennet. The milky sap of fig leaves is a common one, along with a similar latex-rich sap from papaya trees. Nettle has been used in many places. *Galium verum*, commonly known as lady's bedstraw or rennet plant, has a documented history of usage in the United Kingdom. Mexico and parts of the southwestern United States have a history of using the berries of silverleaf nightshade. Exploring the use of local plants as rennet opens another door to terroir and reconnecting cheeses with the landscapes from which they sprout.

4. Salt

As I have traveled and sampled a wide range of cuisines, I have found a nearly universal agreement about how salty food should be. There are cultures known for preferring higher levels of salt, which is often linked to a historical abundance or scarcity of this widely traded, essential commodity. Individual salt needs vary, and people accustomed to high levels of salt expect it, and vice versa. That said, most cultures intuitively understand how much salt to put into food to create the magic balance that satisfies humans. This can be quantified: 1 to 2 percent by weight is what most people consider the proper amount. You can experiment by weighing water in grams, then adding salt at 1 percent of that weight. It tastes to me like a slightly undersalted soup, and this is what I cook pasta in. Meanwhile, 2 percent tastes like a borderline oversalted soup, with a pronounced saline bite. The mouth's experience of saltiness is complicated by the role of other basic tastes. Sweet can soften salt's bite or balance it. Acidity reduces our perception of salty in a more dramatic way, so a very sour cheese like feta may have 2.5 percent salt and still taste balanced. There is a limit, though, and once you go over it, salt overwhelms everything and the mouth is no longer happy.

The only places where I see a tendency to undersalt food is in health-conscious (or crazed) Western countries, where the fear of excessive salt in our diets, perhaps largely due to oversalted processed foods, leads to an obsessive demonizing and cutting out of the ingredient as a health measure. It's in the United States and the United Kingdom that I often feel the need to add salt to food and find undersalted cheese.

A crucial part of cheesemaking that can be tricky to accomplish is getting in the right amount of salt. For most styles, 0.5 to 3 percent is standard, with blues sometimes found on the high end and alpines on the low. Undersalted cheeses come across as lackluster; whatever flavors are there are muted without salt's ability to turn the volume up, bringing the potential into a coherent expression, and sharpening the edges between attributes we can taste.

Salt is the fourth ingredient I discuss because its application is generally at the end of the primary cheesemaking process. The milk has fermented and coagulated; the curd has been separated from the whey and made into its final shape. Then it is salted. Different levels of salt impede the growth of various microbes, so salt plays an important inhibitory role. Many microbes cannot handle 1 to 2 percent salt and will not be able to flourish in a cheese salted to this level. This can select out pathogenic microbes, or those that may cause unwanted outcomes in cheese, such as gas formation, swelling, or undesired flavors.

Salting marks the end of the primary phase of cheesemaking and the beginning of its ripening. Salt can be applied in many ways. Forming whole wheels and then soaking them in a concentrated brine solution for a certain amount of time is common. With smaller and more delicate cheeses, it is normal to apply dry salt directly to the surface. In both brined and surface-salted cheeses, the salt will slowly migrate into the center and establish balance through osmosis; the salt dissolving in the water portion of a cheese and diffusing throughout. Salt can also be added to partially fermented cheese that has been cut up, and stirred to coat these fragments before they are pressed together into a wheel, as is the case with cheddar and most **milled-curd** cheeses. This saves makers the step of brining and allows the salt to spread throughout the cheese faster. In a few cases, this whole process is reversed, with salt being added to milk before coagulation. I have seen this in hot climates, where milk ferments rapidly, and it is desirable to impede the process up front. That said, I have also salted milk at 1.5 percent and made it into lactic cheese that still got very sour. This tells me that the LAB I work with to make cheese are tolerant of this level of salt. The role of salt is probably less about stopping the fermentation that we encourage with starter cultures, and more about preventing the growth of the unwanted microbes that can't tolerate the combination of acid and salt.

Salt can also play a major role in the ripening of cheese and the development of rind ecologies. The final amount of salt in the cheese is a crucial parameter for how it ripens. Not enough and the process could go in strange, hard-to-predict directions, as the playing field is open to more of the riffraff that salt usually prevents from interfering. Too much and you can impede flavor development, preventing the desirable reactions from occurring. When well-made cheeses are adequately fermented and salted, it's rare that anything disastrous or dangerous will happen internally as they ripen. The same goes for the development of the rind ecology. After salting, cheeses exude whey and are obviously very salty on the surface. They are left to dry, and the first microbes that start to colonize this bare soil are invariably yeasts, which can tolerate the salt as they feed on lactic acid and start the growth of a rind. The rinds will go through a predictable succession of microbes as molds begin to grow on the surface that yeasts have prepared for them. Blue, white, and gray are common, and later bright-orange and yellow molds can develop on what are called **natural rinds**, those that don't see much intervention besides flipping and occasional rubbing or brushing.

Salt can be used to steer rind ecologies into a whole other sea of cheeses. Once yeasts have established and consumed some of the lactic acid, bacteria that can't tolerate acid can start to grow. Rinds on which these bacteria are prominent can be encouraged by washing them regularly with water, usually with salt added to create brine at 1 to 5 percent salt. Different salt levels will steer the ecosystem in different directions, selecting for various microbes that are often a part of the raw milk microbiota. With a 1 percent brine washing, you will usually have some yeasts able to survive, while also encouraging salt-loving bacteria that come from various genera to gain a foothold. If the salt in the washing brine is increased to 3 to 5 percent, which most molds and yeasts cannot tolerate, then bacterially dominant communities thrive, and after about two weeks the rinds develop an orange, pink, or red pigmentation accompanied by a robust aroma. Cheeses belonging to this style are often referred to as washed rinds, but I prefer the term **smear rinds** or **smear ripened**. The smear forms as the cheese is kept moist with brine and stored at high humidity. In its early stages, it can be perceived as a slightly tacky, sticky feeling accompanied by the aromatic profile some call stinky and compare to body odor. To me, it often smells like cured meats or overripe fruit, like God's boots after a hard day's work.

As with the cultures and rennet used in cheesemaking, the provenance of salt has become obfuscated—makers tend to simply buy the cheapest salt without concern for its origin or composition. This usually means table salt, which is nearly pure sodium chloride. Like 100 percent chymosin rennet and single-strain starters, salt has been reduced to its simplest form, with any variation or complexity removed from it. It is processed salt, stripped of any individuality or known provenance, the equivalent of pasteurized, homogenized milk. Less refined sea and earth salts often contain various trace minerals; they are more than just sodium chloride. How these complex salts may impact cheese ripening, flavor, and rind ecology is a topic that seems ripe for exploration.

Locally harvested salts could be considered another potential level of terroir in a cheese. In two of the strongholds of regional rennet making I visited, makers proudly work with a local salt despite its higher cost. On La Palma, in Spain, many makers work with a pink sea salt that is produced via wind-enhanced evaporation ponds on the island. These makers feed the salt to a heritage goat breed and use it to preserve the rennet they make from these herds for the cheese of the island. They have such cultural pride in their cheese that they can trace the origins of every ingredient back to its source, which is the island on which they live. In the Basque Country, a cheesemaker (Eneko Goiburu) who makes rennet also works with salt from a brine spring that has been producing salt for centuries. This salt is a part of the heritage and cuisine of the region, and using it adds another layer of local character to the cheese.

How local is a cheese, when three of the four ingredients used to make it come from unknown locations, possibly other continents? The rennet could be from New Zealand, processed in Europe, and shipped to America in a barcoded bottle completely divorced from its origin in the biology of young ruminants. The cultures are made in sealed laboratories we cannot visit in Denmark or France, and the ultimate provenance of the bacterial strains remains undisclosed. The salt is sold in huge bags of standardized table salt, completely severed from its birthplace. Perhaps only the milk comes from nearby, and it may not have much of its potential microbial diversity left due to refrigeration and other mistreatments. This is the situation not just for large industrial-scale cheesemaking, but for the majority of the small, artisan, raw, and farmstead producers as well.

How has cheese become so divorced from the places in which it is made? How have humans become so separate from the acts of living with, milking, breeding, killing, and eating livestock? What are the possible paths to restoring the human-microbial alliances that have led to sustainable farming, healthy landscapes, and the crafting of cheeses that can channel the voice of a place and its people? If cheese has seeds, what are they, and how are they cultivated? These were some of the questions I sought answers to as I headed out for the unmapped trails leading to some of the world's remaining place-based cheeses.

— INTRODUCTION —

Crossing the Threshold

ULAANBAATAR, MONGOLIA

In some sense our ability to open the future will depend not on how well we learn anymore but on how well we are able to unlearn.

—Alan Kay

Sometimes following a bad idea or making a mistake turns out to be exactly what we need to get us pointed in a positive new direction. Our rational minds might see potential danger and failure ahead and warn us not to pursue an opportunity, passion, or relationship. The internal worry committee tells us not to go down the confusion of paths leading into the forest of the unknown. But there are other voices, other selves and forces, prompting us to follow our hearts, no matter what hardship or hypothetical roadblocks may lie ahead. The unmapped trail leading into dark thickets is scary but also alluring, and one of infinite possibility. Such was my experience in taking a job offer to manage a cheese plant in Mongolia. My fascination with the country had begun years prior, when I read a book about Mongolian cosmology and the wind horse, a divine power that allows us to accomplish great tasks effortlessly, that infuses existence and carries our prayers to the heavens. If I had never picked up that book and envisioned myself riding a horse on the steppes, it never could have happened.

Life's untamable adventure isn't found by taking the easy route where you can never get lost, scrape your shins, or takes steps you later regret. We learn to walk by falling down and standing back up, over and over again, stumbling, bumbling, and eventually running in the process of finding our feet, our balance. Hopefully, after years of tripping, we can learn to embody the graceful ways of the wise, who can fall down, grieve, and go mad beautifully, who can make mistakes and take wrong turns with composure, who realize that sometimes losing the trail *is* the trail. Seen from the eagle-eyed vantage of the soul's flight, untethered to earth's supposedly linear time, there are no missteps. Only the walking of the path.

Many red flags waved in my mind's eye; I feared that I wouldn't enjoy working for this company, that it was built on shaky foundations. I took the job anyway. It was a bad move, and the best decision I ever made. Taking this job that I would soon quit led me to traveling in the Mongolian countryside, my first steps on a path that would take me out of the valley of grief to passes between peaks of hope and optimism for the future of our confused and troubled species. If I had listened to the hesitations of my rational mind, if I had been afraid of taking a wrong turn, I never would have jumped into the adventure that has unfolded so beautifully and bumpily over the last five years.

Gazing out the window of the plane, my eyes scrolled across a flat, unfamiliar landscape. I looked down at the Gobi: brown, treeless rock gardens and sand. An occasional settlement at the center of spokes of automobile tracks radiating out into arid country. A lonely highway, scars of mines and train tracks, small ranges of hills and buttes. The human artifacts increased and coalesced into Mongolia's capital city, Ulaanbaatar. The air on the tarmac had a strange, high-elevation, New Mexico quality to it, a roof-of-the-world feeling: dry and cold, with a thin atmosphere saturated with sunlight. A visible haze of smog spread across the valley floor, from the thousands of homes where coal was being burned for heating and cooking. I was met by two men who would be my co-workers, and we rolled through a beat, former state-run factory town called Bio Kombinat, a suburb of UB. Shabby dogs slinked between decaying Russian-era block apartments as we crossed a rickety wooden bridge to the base of a small mountain range alongside a frozen river where trash and vodka bottles were trapped in the ice, waiting out the winter.

I settled into the ger that I would live in next to the cheese plant. Gers are the circular structures known to many as yurts, a Russian word. They can be packed down in a few hours and put on the back of a truck with everything a small family owns. The ideal house for mobile, semi-nomadic cheesemakers, a lifestyle I was captivated by and aspiring to. Moving my life from the giant, toxic box buildings of North America to this circular structure with colorful, sacred knot patterns hand-painted on the doors, I was crossing a threshold, leaving the familiar world of straight lines and fixed boundaries for a more permeable, wavy, spherical reality.

My job involved developing cheese recipes and creamery protocols while training Mongolians for the other plants the company was setting up in a network around the country. An awareness crept into my mind that I actually had a great deal to learn from these Mongolians whom I was supposed to train, that their cultural knowledge of milk and its chemistry was very advanced and contained an ontology and form of wisdom that diverged from the worldview in which I had been steeped. I watched them do things like dip their fingers in the milk before we made cheese and flick it in the four cardinal directions. A similar phenomenon can be seen every morning around the country, where women greet the sun by throwing milk or milk tea into the air as an offering while repeating prayers. I sensed that, to Mongolians, milk is sacred, that every part of it is precious. That milk is a gift from gods and spiritual forces that my culture had told me were dead. This mythological framework intuitively made sense to me. It felt like a more functional and healthy reality to live in compared with the mechanical universe of so-called rational laws and objectivity that is erasing our humanity and trying to take the rest of the planet down with Western civilization's sinking ship.

My co-workers often appeared visibly upset when we put whey down the drain, as if we were casually dumping holy water onto the floor. It is called *shar suu* in Mongolian, which means "yellow milk." It is still milk, with the cheese taken out. Up to this point, I had thought of whey as a byproduct of cheesemaking, a constant nuisance, a breeding ground for bacteria. In Mongolia, whey is used to make a ricotta-like cheese, to cook into soups and grains, or to bathe in. It is reputed to be especially good for washing one's hair and is valued as a beverage for people who have digestive troubles. To consider it a byproduct, something to be disposed of, is a sacrilege.

It was morally conflicting to receive fresh milk directly from herders, still warm from the bodies of animals eating grass, which we immediately pasteurized and cultured with commercial starters. We were making cheeses modeled on those of Europe: camembert, cheddar, an alpine **tomme**, a blue. It felt colonial on many levels, this imposition of an industrial model and Eurocentric vision of what cheese should be. To see Mongolians treat milk with such reverence, which we then desecrated by adding disposable packages of freeze-dried microbial imperialism, pained me.

Wanting to learn about the traditional cheeses of Mongolia, I had the sudden and convincing realization that I needed to quit this job and start walking, traveling lightly with just a backpack. To visit people who lived *with* livestock, *within* landscapes, milking and making food on a domestic scale. To see what remained of older, more practical approaches to making cheese that are not reliant on purchased inputs. To learn the techniques and recipes before they were buried in modernity's rush to slaughter everything slow and holy. An enchanting melody drifted in on the wind, from the vast, peopled steppes. An irresistible song whose strange instruments were somehow familiar. It was the voice of the earth herself, calling me home.

A Reckless Omnivore Is Born

From a young age, I've carried a sense of loss and grief over the destructive path of my culture. At six, I found joy and a sense of completeness in forest walks with my sisters. The only spirituality we knew was that of respecting Mother Nature. Then, on a car ride home from kindergarten, my heart was torn with concern as my mother explained to me the existence of endangered species. I was stricken with a sense of tragedy, and with anger at the world I lived in for causing this loss—emotions I was not equipped to process. As a teen, my grief accumulated as I realized that the culture I inhabited was based on material gain, and all the life paths it offered were shallow and subservient to a structure built on a rotten foundation. Something was horribly wrong; the tune everyone was dancing to was discordant, and the syncopation drove me to despair. I rebelled, became a skater, and read about the fall of civilizations, with the conviction that I had been born into a collapsing global empire.

This conviction has only grown stronger in intervening years. In college, I studied anthropology and philosophy, following a fascination with preindustrial lifeways, tribal societies, and the negative impacts of the agricultural revolution. I was introduced to an intriguing notion: that the root of our social and political problems was in how we related to the land, how we fed ourselves. My grief and anger were composting into fertile soil from which saplings would grow, promising future fruits to feed my soul, sweet offerings to place on the neglected shrines to holy nature that I find and refurbish everywhere I go.

My passion for food and cooking emerged after being an extremely picky eater as a child with food allergies (cow's milk and citric acid) and a sensitive, alarmist palate. My mom had me drink goat milk and slowly introduced minute amounts of citrus, and eventually my allergies faded.

Despite her efforts, as soon as I was old enough to rebel, I refused to try new foods and stuck with the peanut-butter-and-jelly sandwiches and Top Ramen I prepared for myself. I was skeptical of other people's cooking, suspicious they would expose me to novel, offensive flavors. I knew what I liked, and I was empowered by learning to cook bacon, scrambled eggs, toast, and macaroni and cheese. Eventually I began choosing recipes from old cookbooks. My mother would buy all the ingredients and walk me through the techniques. My palate slowly opened, but certain types of food, especially seafood, disgusted me; the smell of canned tuna made me gag.

One evening, while eating dinner at a friend's house, a huge fillet of freshly caught salmon was praised by the whole family. They refused to accept my denial of this holy, wild food, and insisted I try a bite, with lots of lemon juice. My heart raced in fear and I nearly ran home crying. I wish I could go back and see the look on my face in that moment as my horrified frown-grimace of anticipated disgust flipped into a smile of revelation at that first bite. Not only was it not disgusting; it was delicious! It didn't taste fishy at all, and I was pleasantly shocked to find myself digging in. The picky eater was pushed aside by a new kid on the block of my consciousness. I liked this new kid; he took risks, knew how to have fun. Upon returning home, I told my mom, "I ate salmon, and I liked it!" All it took was visiting someone else's kitchen and a bit of peer pressure for the walls of my gastronomic prison to start cracking, for the liberating light of palate expansion to shine in and illuminate a new path.

This began a practice of challenging my ingrained predilections. I found that growth was accelerated by exposure to novel foods and ideas, by sitting through the discomfort of the unfamiliar, eating in someone else's home, sleeping outdoors in the cold. What attracts or repels us about food is culturally and psychologically determined. It's less about the actual sensations in our mouths and more about context and associations, how we categorize and relate to foods. Our cultures and families teach us what is and is not food, what does or does not smell good, which textures are pleasant or unpleasant. Whatever Grandma feeds you is good, even if it's a cheese containing living maggots (*casu marzu*), fish preserved in lye (lutefisk), a boiled egg containing a partially developed duck fetus (balut), deep-fried insects, or Cool Ranch Doritos. I've had all these foods. Some of them smelled and looked extremely unappealing to me, and it took summoning courage or a few drinks to overcome my preconceptions and pop them in my mouth. I didn't always enjoy them, but I am glad I at least tried, because my culinary frame of reference is now wider. We can learn to like foods if we set that as an intention. Not by trying everything once, but by trying something over and over again, until it makes sense or we acquire a taste for what we formerly dismissed as disgusting.

During my first international backpacking trip in Southeast Asia, when I was thirty years old, this practice became fixed as a key aspect of how I would travel and live. I had never enjoyed cucumbers. They were boring: dirt-flavored, limp-crispy water chips. Not gross, but vaguely annoying and in the way, getting vegetable slime all over an otherwise pleasant salad. In Cambodia, almost every meal—spicy noodle soups, pork on rice—came with sliced cucumbers on the side. Since I was trying to spend as little money as possible, I cleaned any plate put in front of me, including the ever-present cukes. These were different, much more flavorful than the bloated green water logs sold in American grocery stores. These varietals were bred to actually taste good, to express cucumber chillness rather than be a mass-marketed commodity that ships well or is easier to grow as a monocrop. They paired well with fiery fare and supplied textural diversity while serving as a cool, refreshing counterbalance to the tropical heat. I finally understood their role. It just took traveling somewhere that had good varieties, and exposing myself to the cuisine of a culture that knew how to use them. It took eating them repeatedly while wanting to like them until I enjoyed and even craved cucumbers.

The realization struck me that I could do this with any food. I began seeking out the most challenging dishes, ones that my former self would have rejected as disgusting. Dried fish, organs, beans fermented with mold; every fruit, vegetable, egg, or animal became fair game. This was a way of exploring culture, exposing myself to it by eating in restaurants crowded with local people, ordering what they ate, taking food tours, and learning how people live by cooking with them, staying in their homes, listening, watching, smelling, tasting.

A Seed Is Planted

Driven by this anthropological fascination with gastronomy, I spent time cooking in restaurants, pursuing a newfound love of fermentation. I had no conscious intention to become a cheesemaker. Answering a Craigslist posting, I got my first cheesemaking job in downtown Seattle. At this job,

I was exposed to precisely the approach that I would eventually be opposed to: pasteurization, commercial starter cultures, chemical sanitation, a mentality of mastering microbes. We toiled like overworked mules in an urban setting far removed from barns and pastures, and I never saw the cows who gave us milk.

One afternoon, while drinking beer in a dimly lit bar full of pinball machines and apathy, my friend and co-worker Dan Utano laid out for me how raw milk cheese can have terroir. *Terroir* is a French term commonly used to describe how the character of a wine can be specific to a place, rooted in the soil, geology, weather, cultivation practices, and fermentation techniques. Its linguistic origins are in the word *terre*, meaning "land," "earth," or "soil." There is debate over whether the term is applicable to cheese or not, with some seeing milk as a universal substance whose origins have a negligible impact on a finished cheese. Dan explained that terroir can manifest in raw milk cheeses because microbes living on the plants that livestock eat and in the barns where they sleep can transfer into the milk and influence its fermentation. Cheese can reflect the place it is made, as a product of a landscape. I wouldn't have a tangible experience of terroir for another decade, but the possibility of it struck a chord deep inside me. I wasn't conscious that this would become the thing I would fixate on, but a seed had been planted in my heart.

This seed sat dormant as I spent the next nine years making cheese commercially around the United States. From Seattle, I moved to a sheep dairy in California, where I made sheep milk cheeses for two years and fell in love with the craft and with farm living, being close to livestock, plants, and soil. Raw milk cheesemaking on a farmstead level addressed many of the ills of the industrial model. I saw the sheep on pasture and transformed their milk into cheese, which I cared for while it aged before wrapping it for sale at farmers markets. For the first time in my life, I was participating in work that I believed in, that was more than a job. I was finally taking action, focusing on a positive response to my childhood grief over species loss and environmental degradation, channeling my rebellious teenage angst, and evolving philosophically as a young adult. Rather than fixating on what I was against, I could now direct my energy toward what I stood for, toward ways of addressing or walking away from the broken system I had spent years studying from the inside while feeling stuck there. Flipping the script changed everything. I was filled with passion and drive, expanding my knowledge of milk's fermentation by reading, asking questions, and repeating the daily work of turning milk into cheese. Cheese became something that liberated and fed my soul; it opened doors, allowing the wind horse to blow through and carry me away.

It felt radical, like I was participating in revolutionary alternatives to the conventional models of agriculture, of cheesemaking, of structuring my life. I did not yet know that the industrial mindset of making cheese had been imposed upon small-scale producers, too. I hadn't learned that things like the refrigeration of milk, instead of safe and necessary, were in fact damaging to raw milk, led to the loss of its healthy microbial ecology, and made it a less safe food. I couldn't see that raw milk and farmsteads were not enough, that any potential for terroir was being overwhelmed by packaged cultures, stamped out in overly sanitized environments, neutralized by a neurotic anti-life mentality that disguises itself as science-based food safety. I spent years bouncing around—Colorado, Vermont, back to California—working as a full-time employee for companies of various scales, moving to wherever I found a job. I loved the work, but there was always a phantom-limb sense of missing something important. So I kept moving, searching for the pieces of the puzzle that I could not yet identify.

I had internalized elements of the industrial approach. In my mind, delicious cheese could be crafted by following rigid protocols in milking and cheesemaking. It was about blending the packaged starters in a particular way, tightening a recipe, and precisely measuring variables like acidity and moisture. Through attentive monitoring and record keeping, with strict control of conditions in the creamery and aging space, cheese rinds could be cultivated from their own packaged cultures of mold. At this point, I didn't know how mediocre the cheese I was eating was, didn't realize that even the imported raw milk cheeses from Europe were made using the same standardized cultures and rennet used by artisan producers in the United States. I had never tasted cheese made any other way. I didn't realize there were alternatives. Having never tasted anything wild or expressive of terroir, I simply didn't know what it was.

After reading a subversive book called *The Art of Natural Cheesemaking* by David Asher, I realized there were whole other approaches to making cheese and that the use of packaged starter cultures that is so dominant, from large companies down to home-scale cheesemaking, was definitely not how I wanted to ferment milk. I began questioning the whole paradigm in which I had been trained. Were the microbes needed to make cheese already in the milk? Were the packages of culture and bottled rennet produced by corporations on other continents actually moving a cheese farther from the place it was made? Like that conversation about terroir in a Seattle bar, the concepts of natural cheesemaking fueled my fire.

In hindsight, it is clear that to comprehend what cheese could be, I would have to go taste as much of it as possible. I would have to seek out and eat in place the rare and endangered cheeses that don't travel far and are often made outside draconian food safety regulations. In taking the job in Mongolia, I had no conscious intent to turn my whole life in this direction. I had no ten-year plan. I simply had to go. Something outside my everyday conscious mind was pushing me, allowing other forces to pull me forward, onto the frozen steppes of novelty. All I had to do was put one foot in front of the other, falling forward into life's loving hands, letting the wind horse carry me away.

Pastoral Cheese

When I wasn't on the clock at the cheese plant in Ulaanbaatar, I spent the cold winter days and long, starry nights learning as much as possible about Mongolian dairy foods by reading and talking with friends and co-workers. At a large market in the city, I was able to sample some of the foods I read about, stacked in bins as large piles of very dry, hard cheeses next to mounds of butter cut with vegetable oil. Winter suddenly surrendered to summer in May, with no intervening spring. I walked the short distance from the cheese plant into the rough village set at the base of Wild Onion Mountain. Many of the houses did not have electricity or running water, and there were gers set up next to ramshackle homes hastily assembled from repurposed materials. Poverty and alcoholism were apparent. I would attempt to ask anyone I saw if they knew of people milking goats by saying "*Yama suu?*" (goat milk) and performing the universal hand gesture for milking.

A group of laughing children adopted me into their gang, and they walked me to the home of a friendly woman who invited us in for lunch. Using Google Translate to approximate my questions into Russian, she replied by combining body language and bits of common English, explaining that the baby sheep and goats drink their mother's milk for the first two months of life; the herders start milking regularly in June. This marks the start of the famous "one hundred days of summer" during which the grasslands grow quickly and the herding families of Mongolia are busy storing away the bounty of grass-fed milk as various foods. The short summer causes an explosion of photosynthesis, and this energy is transmitted by livestock into meat, fiber, and milk. The milk is preserved as dry cheese and sour alcoholic mare's milk to provide calories throughout the intervening 265 days of colder weather.

Cheese in Mongolia is without the pretensions of fancy food, wine pairings, and the artfully decorated grazing boards that are associated with the food in the United States. Milk and cheese are a cornerstone of a time-tested sustenance strategy based on the raising of ruminants, likely one of the only realistic and sustainable approaches for lands far from the equator—those with little arable land and a continental climate marked by long, cold winters and short summers. Crop agriculture has never been a viable means of feeding people here. It is through integration with and symbiosis between large herbivores and grasslands that Mongolians have found a way to thrive. Plants are transformed into milk and dung, then dried dung is used as fuel to cook milk into cheese. These cheeses are not what we commonly associate with the word. There are no moldy white rinds, wax-coated wheels, blue cheeses, or huge blocks of cheddar. Aging in dank caves is not an option; semi-nomadic pastoralists don't have cellars, and this is an arid climate of extreme temperatures. The main route toward preservation is to divide milk into cream and skim, then dehydrate it as a range of foods such as yogurt, ricotta, and a paneer-like cheese. This is cheesemaking in its essence, stripped of snobbery, done without the products deemed necessary by the world's highly esteemed cheese-producing nations.

My first hands-on experience with these processes, which I had only been absorbing intellectually, came while I was driving with the cheese plant's engineer and interpreter, Enkhbat. While he navigated a maze of dirt tracks, I asked if any of the ger camps we passed were processing milk. "Where there are herders, there is cheese," Enkhbat told me in the concise, deadpan, yet poetic Mongolian manner of speaking. "Where there is smoke from the chimney, there is milk tea." We drove up to a ger that had the typical array outside: a satellite dish, a motorcycle, solar panels. Large, scruffy dogs came running, flexing their muscles while barking protectively. Enkhbat called out what I knew was a common Mongolian greeting that translates as "Call off your dogs." A man emerged from the ger and yelled sharply, "*Haaaat!*" The dogs ceased their barking and cowered, shifting their body language from aggression to skepticism. A short conference was held between Enkhbat and the herder as we stood in the vicious, unrelenting steppe wind, which proved more difficult to placate than the dogs.

Upon entering the ger, I was greeted by a scene that felt contrived to please me. A woman sat at the central wood-burning stove, the same seen all over Mongolia, where fire is deemed holy. Sunlight shone down from the skylight. It felt like a cathedral, except pots and buckets full of milk and cheese were everywhere. The woman was slowly stirring milk and began fussing with the food store behind her. There is a prescribed layout inside a ger, with female and male sides, and a family shrine directly opposite the main door. Beds, which double as couches, have their assigned locations; meat and cheese are hung to the left of the door, and the pantry and cooking supplies are stored on the right. As soon as we sat, this woman began covering a small table with plates and bowls of milk foods. Enkhbat described what each one was and explained that the woman was now processing the whey left over from one cheese into the Mongolian take on ricotta. Ricotta is made by heating the sweet whey resulting from a rennet-coagulated cheese until a delicate curd appears around 180–185°F (82–85°C), sometimes with the help of a small dose of acid. This is ladled out and drained, to be eaten fresh. When the whey was steaming, the woman stopped stirring and called us over to look. I saw the familiar formation of what I would call *ricotta*, an Italian word meaning "recooked" or "reboiled." (Rather than using the cumbersome phrase *denatured albumin whey cheese*, I will just use *ricotta* as shorthand for the various analogs of this technique.)

There is no one, universal whey. There are many wheys. The whey resulting from a **rennet curd** can be more or less milky, depending on how much fat was released by cutting and stirring. The wheys from goat, sheep, and cow milks are different, and in Italy there are different approaches to turning each of these into ricotta. The whey from a long-fermented lactic cheese like chèvre or yogurt is very sour, in contrast with the sweet whey obtained after making a rennet curd. There is the whey left after ricotta is made, which is called ***scotta*** in Italian. Many milk foods—such as whey, feta, or sour cream—turn out to have variations and deep nuance that are hard to appreciate until you visit a place where this diversity is expressed.

The woman's process now deviated from the ricotta approach. Rather than scooping out the delicate curds, she removed the scotta, leaving about two inches in the pot with the ricotta. This was allowed to simmer and reduce for over an hour until it turned to a thick syrup that coated the curds, the combination beginning to brown and smell of caramel. The resulting sweet, firm ricotta called *eezgi* was spread on boards and placed on the roof of the ger to dry in the sun and wind, becoming firm, chewy granules that could be stored long-term. It was also formed into fragile cakes with a texture similar to shredded coconut. I imagined the ricotta was kind of mashed up before or during drying to give it this texture. Eezgi is one of my favorite Mongolian cheeses, and I have successfully approximated it while teaching workshops in warm, dry climates. The technique could likely be replicated anywhere with an analogous climate, or by using a dehydrator or oven.

I had simply shown up as an unannounced stranger and been allowed in to observe the domestic life of this family who answered my pushy questions about these processes that were mundane to them. All around this country, similar scenes occur daily in the summer: the laborious taking apart of milk, cooking it over a fire, making cheese in a very self-contained manner that requires zero inputs besides metal pots, stirring implements, draining cloths, and drying boards. No packaged starters. No bottles of rennet. No triple sink with alkaline, acid, and sanitizer to maintain pristine stainless-steel and plastic "food safe" equipment. Milk is processed directly, with no intervening refrigeration, in a manner that has been practiced for a very long time—hundreds if not thousands of years.

There was a clue here to what had been missing all those years I spent making cheese in the United States. I was compelled to see more of this dairy sovereignty, to learn by observing and participating, taking notes, photographs, and videos to document this deep knowledge and share it with a world that was hungry for these stories, hungry for reconnection with landscapes and one another through real food. These remaining pastoral people could help us reimagine what it is like to live with livestock, fermenting milk, participating in the cycles of nature as active members.

I believe that traces of an older, saner, time-tested wisdom can still be found in all human hearts and cultures, even if they are deeply buried and visible only in small pockets. The essence of our true selves—our dignified, indigenous, human souls—survives. I've seen it, felt it, tasted it. It was in Mongolia that I first stumbled upon that faint trail of cheese crumbs, leading into the hills and out onto the steppes. The call to adventure was becoming comprehensible, and the concept of milk trekking was born.

— CHAPTER ONE —

Eleven Days in the Valley of the Yaks

ZAVKHAN PROVINCE, MONGOLIA

The spirit of the valley never dies.

—Tao Te Ching

In the ger behind the cheese plant, I reflected in solitude and pulled myself together, building strength for the journey ahead. The rocky peaks a short walk away were my therapist; I told them everything and left offerings at rock-pile shrines located on the summits, where others had left tobacco, tea, pastries, and prayer flags. I meditated, wrote, repeated mantras such as "Every thought sends a ripple into the future" and "I am rewiring my neurons." Positive ripples were sent, neural networks rewired. Listening to talks by Tibetan Buddhist philosophers returned truths that had become dusty in my mental filing cabinet to the front of my consciousness. There was nothing to fear in the entire universe: I could stand up and walk, letting my feet choose the path, finding goodness in any direction.

In July, after working at the cheese plant for five months, I quit the job and left the riverside below Wild Onion Mountain. In order to get my passport stamped with a tourist visa, I took the train across the Gobi to cross the border with China, then immediately reentered Mongolia. On the overnight train back to Ulaanbaatar, I met a young Mongolian woman named Burmaa who spoke fluent English as well as Chinese. She wore her long black hair in a braid. Her smile started in one corner of her mouth before spreading across her face, and her demeanor switched between playful and stoic. We drank beer on the train and had our passports confiscated by a power-tripping train cop who seemed offended to see me getting friendly (and maybe a bit drunk) with a Mongolian woman. I remembered someone had told me that "All Mongolian men think they are the descendants of Genghis Khan." Of course, many of them actually are. The old Soviet-era train chugged into the city, our passports were returned, and I met up with Burmaa again a few days later. When I told her of my desire to spend time with herders in the countryside, she invited me to come stay with her family, who herd yaks in Zavkhan province. A week later, I was on a bus heading to the west of the country, passing through the flat steppes and heading up into forested mountains that resembled Montana.

Mongolia is twice the size of Texas, with a population of only three million. Some estimate there are twenty times more livestock than people in the country. A diversity of dairy animals are kept: cows, sheep, and goats are milked, as well as camels, yaks, horses, and, in a small pocket, reindeer. I wanted to learn about the milk of a less common species, and yaks held a particular appeal. As bovines, their milk shares characteristics with cow milk: a large **milk fat globule (MFG)** leads to cream rising, the milk can contain carotene lending a yellow hue, and it readily stretches into mozzarella. Yaks tend to be much smaller than cattle but their thick coats of fur make them appear larger, like fluffy dogs. Their hooves are smaller, so they impact landscapes with a lighter hoofprint. Yak fiber is being offered as an alternative to the cashmere of goats, whose grazing is blamed for the degradation of grasslands in Mongolia. Yaks are adapted to a particular landscape: high-elevation steppes and rocky, mountainous rangelands. Unlike many modern cattle breeds, yaks have not been bred to produce large amounts of milk, something that requires high-caloric diets that rely on grain supplements. They still possess hardiness and can thrive while grazing on meager forage. Something less submissive and tame, more solid and regal seemed to ripple out as they ran across the unfenced earth nurtured by their existence.

We spent a few nights in the regional capital, sleeping on the floor of Burmaa's mother's small house, then paid a driver to take us out in a Prius, the most popular car in Mongolia, which you see in the most improbably rugged terrain. We left the pavement for good and drove for hours on unmaintained roads that traverse "the countryside." The term seems to denote 99 percent of the country, which is rural space with few power lines or settlements, where families of herders live. Some of it is what many would call wilderness: uninhabited forest, mountain, or desert zones, or grasslands that are periodically grazed. As we crested a pass along a rocky ridge, a vista opened on a verdant valley with a river snaking through unimpeded, past small herder camps with a few gers and vehicles, with huge multicolored herds of sheep and goats roaming in between. Human habitations and reordering of the land and water were not arrogantly imposed but followed a more organic, respectful pattern.

Driving down the valley, I noticed that each herder camp was separated from the next by a buffer of a half to three-quarters of a mile. Burmaa explained that each year the same herders return to this valley, their families having some sort of customary usage rights. As the families arrive, they simply take the next reasonable location up the valley, having no claim to a specific piece of property. The buffer zone keeps their herds relatively separate to avoid the chaotic mixing and subsequent laborious sorting that can occur in this unfenced valley populated by a dozen families and thousands of animals. This introduced me to the esoteric land-use policies of pastoral peoples, which are often noncodified and hard to grasp by a mind trained to divide land into simplistic, anthropocentric categories of private versus public. There were definite rules and norms, and consequences for breaking with these, but they seemed to belong less to the state and more to the realm of custom and tradition.

Burmaa's sister and her young sons came out to greet us as we drove up. The dogs barked, smelled us suspiciously, and calmed down. We ate a big lunch inside the ger around a short table. The two boys would be my friends and teachers. Their father had gone to the city for surgery, and the elder, who was probably eleven years old, took his new role as man of the house seriously, chopping wood and carrying water with gusto and a poker face. They saddled up an old horse, of the small and sturdy breed common in Mongolia. Horses were referred to by physical features rather than names, so they called him "beige with blond locks and a dark muzzle," which was one long word in Mongolian. They invited or rather dared me to ride the stubborn, patient, slow animal that I nicknamed Old Bill. I dug my heels in but could never get Old Bill to travel at a speed above a slow walk. It was a comical sight, a tall white man atop a small horse. The boys couldn't believe that I didn't know how to ride, this being like riding a bicycle in Mongolia. We all laughed. Even Old Bill chuckled a bit. When you find yourself in an unfamiliar land where you don't speak the language, become a clown and laugh at your predicament, at the awkward beauty of your blessed life. Linguistic barriers are not so tall when we apply the universal language of laughter, the common speech of utterances, body language, and playful mimicry.

The boys took off on Old Bill and a bicycle as the sun got low, riding out to collect the yaks who were grazing in a valley beyond sight. The herd of sixty or so came bounding in majestically, hair flying in a blurry, visionary dream of mythical bovines painted on the wall of a cave. Upon arriving at camp, the calves were caught and tethered to the cold ground of reality. They wore collars woven from yak fiber with a wooden toggle hanging down like an amulet. This short piece of wood slid through a loop at the end of a short lead, securing the calves to a longer line that was staked to the ground on both ends. This long line was rotated a quarter turn every few days, creating a circle of heavily impacted soil with much manure and urine. It was then moved to a fresh spot, spreading out the impact caused by the herd bedding down for the evening. With the babies tethered and lying down to nap, the moms would stay close, munching on grass or drinking from the river that was about a hundred yards from camp.

The Daily Work of Herding

Close to sunset, Burmaa and her sister came out wearing the long, colorful robes known as *deels*, a famous item of Mongolian dress. Some are still made of felted animal hair (sheep, yak, camel), but more commonly they are made from the synthetic materials whose invasion of cheapness is resulting in the steady loss of animal-fiber processing and its artistry in much of world. Burmaa usually wore fashionable Western clothes and makeup and seemed to be an urbanite, so I was impressed to see her don the deel and wrangle up livestock.

The mother yaks were gathered up again with minimal prodding, as they understood the routine and returned to where their calves were tethered. Many of the mothers were hobbled by having their front legs tied together. The babies were released one at a time, running straight to mom to begin suckling, tails wagging furiously. After a few seconds on each teat, the babies were pulled off and tethered to the long line again, or directly to their mother's neck collar. One of the women would then sit on a stool alongside the mama yak and milk her out into a pail. Some mothers allowed this to take place peacefully while chewing cud, not even requiring hobbling. Others comically tried to hop away while their legs were tied together. One of the boys would lean into their necks, standing in between the yak and the woman who was milking. This visual barrier seemed to trick the mother into thinking it was the baby continuing to suckle, keeping her calm so the milk could flow.

With some older livestock breeds, the babies must suckle to stimulate the flow of oxytocin in the mother, causing milk let-down, which is the release of milk from the mammary gland ductwork into the teat. By allowing the babies to suckle, the most natural way of cleaning the teats is accomplished. Saliva likely has a positive impact on the microbiota of the teats and, in turn, the milk. There is also a profound feedback loop that happens between the mother and her baby, with the mother's immune system responding to cues from the baby's saliva. The components of her milk can actually shift to address health conditions in her child, such as increasing antibodies specific to an illness in the young animal. If animals are allowed to suckle, they will have the best likelihood of growing up strong and healthy.

After evening milking, the mothers would be tethered to individual stakes away from their calves. This allowed their udders to fill with milk overnight. In the morning, the calves would drink from their moms momentarily, and then humans would step in and milk the yaks out. Then the calves were allowed to spend the day with their moms, as the whole herd got pushed away from camp by herders on horseback or motorcycle. The yaks formed a social entity distinct from the much larger herd of cashmere goats and sheep, who numbered perhaps 350 to 400. This group of small ruminants also spent the night close to the ger camp, usually coming in of their own volition. I was told they did this because they understood the danger of wolves and desired the salt blocks kept near the gers. At other times they had to be gathered up, especially if they had crossed the river. They would have to be forced back across, which was difficult as goats notoriously avoid getting wet. I joined the men in cajoling the herd to the river's edge, physically pushing a few of the leading animals into the water. Once these first few had crossed, the rest started to follow, with reluctant backtrackers being urged to follow the herd by our waving hands and loud utterances.

There was a single man in his early forties who lived in a ger in the same camp as Burmaa's family and worked closely with them. One day I joined him on horseback, moving the large herd of goats and sheep up into a surrounding valley, then keeping an eye on them as they grazed. Old Bill the curmudgeonly horse did well, seeming to follow the herd wherever they went without any direction from me, the green horseman. I could smell the condescension as he scoffed and rolled his eyes at me when I attempted to guide him. The very way he breathed seemed to say, *Just let me handle this, city slicker; you're out of your element.* He would barely even look at me and acted all put-out when he had to. I liked him immensely and aspire to someday be so comfortably grumpy and over it.

We rode to a ridge above the herd and sat in the shade of some rocks to eat lunch as the animals spread out like raisins in a fruitcake and settled into peaceful grazing. As we relaxed there, the man began to sing a beautiful and eerie song. The words I could not understand, but the power of the song and his voice was conveyed to my core, giving me goosebumps. We napped long, waking from time to time to sit up and see how far the goats had wandered. We watched as another herd of goats crested the opposite hill and began descending into the valley. The man began to whistle and yell excitedly, I assumed in an attempt to communicate with the intruding herd. Soon a herder on horseback appeared nonchalantly on the hill, and an exchange of friendly hollering ensued. This mounted herder quickly rode around his animals, pushing them back up the hill and away from ours, which satisfied my man, who lay back on one elbow, smugly chewing a blade of grass. I can't remember what we ate, but I will never forget the feeling in that song. It felt like memories rapidly recovered from amnesia, a download of what is true and at the core of humanity. It sounded like goodness and grace and truths that can never be uttered in mere words. It sounded like the dignity we are all capable of, the undefeatable decency at the core of the human heart, which the forces of greed and war can never conquer. It sounded like the earth itself, channeled through one body into another, hinting at the unity underneath our skin.

Time spent in the company of shepherds without a shared spoken language has shown me that more subtle languages exist, which are common to all humanity. So much of what is important is conveyed through facial expressions, inflection, volume, and the energy behind the words. There is also a very tangible energetic exchange that happens beyond the grasp of the five senses and takes place simply by being close to others, when our fields overlap, provided we don't have our shields up. There are only so many situations humans can get themselves into, and the context combined with how the words are spoken betrays much of what is behind the sentences. When these herders yelled to one another, I could guess what they were likely saying. When a mother scolds her child, you don't have to know the language to understand the message. The vibe conveys as much meaning as the words. Sometimes it contains more.

Butchering Day

I managed to launch a barrage of questions at the people who lived seasonally in this valley, working with Burmaa as my accomplice and translator. At one point I asked if there was any knowledge of rennet being used to make cheese here. They asked me to explain what rennet is, and as I did they scrunched their faces in disapproval. The response was, "We would never kill a healthy young animal just to extract this organ. You obtain much more meat by raising it to full size. We try to keep as many animals alive as possible, since our families are as wealthy as the number of animals we own." The technology of rennet was being used in modern dairies, but these knowledge holders saw it as completely unnecessary. They had a range of time-tested approaches to preserving milk that did not involve any rennet whatsoever.

It wasn't that there was an ethical issue with the killing and eating of livestock. Meat is a huge part of the Mongolian diet, and here it was of the highest quality, with the feeding of grain or hay grown with chemicals being rare. The meat of heritage-breed animals feeding on native pastures, drinking clean water, and breathing unpolluted air is the kind we would pay a premium price for in the United States, and it is standard fare in the Mongolian countryside. I watched in astonishment as a sheep was killed by rolling it on its back and making a small incision just below its sternum. Reaching into the body of the still live but calm animal, a hand maneuvered to the spine where an artery was severed by expert fingertips. The animal slowly drifted out of consciousness as the cavity of its rib cage filled with blood. This was carefully ladled out into a bowl that was stirred rapidly to prevent it from coagulating. It is taboo to let blood hit the ground in Mongolia, so it all must be caught and consumed. Besides, as I would soon discover, blood is delicious and nourishing. We should all be consuming the blood of the animals we kill, in order to honor their sacrifice more fully. Blood, like all nectars of the body, is sacred.

The hide was removed, and the organs given to the women, who cleaned them out and poured blood mixed with wild onions into the intestines and stomach compartments. The animal was quartered quickly and methodically with a hatchet and pocketknife by the elder son, who had sharpened them meticulously on a rock beforehand. Butchering is a male task, and this young man had obviously learned it well. That night, we feasted on just the organs, all cooked together in a huge pot. Pieces were taken out one at a time once deemed done, a status that everyone argued over, as people do wherever meat is cooked in a social setting. We all sat on the floor and ate out of a communal dish, Burmaa telling me what each piece was. I had never been a huge fan of organs, but this meal helped me acquire the taste. The blood sausage in intestines was lovely. The lung less so. Organs and blood are like seafood, I realized: freshness is everything. These parts are seen as waste by a world that views the time and effort spent cleaning and preparing them to be not worth the trouble. This is an extreme inversion of values compared with the many societies where it is the organs and blood that are considered the most valuable parts. Organs and blood are nutritional powerhouses that also spoil fast and so should be consumed first, while the meat hangs.

With no refrigeration, meat could only be kept cold in the winter months. The sheep we had butchered was eaten within a week. The hanging meat stopped dripping blood and developed a thin, tacky-dry, lacquered outer coating. After the whole animal was eaten and shared with neighbors, the last few pieces were kept under a mesh net to prevent flies from getting at them. This failed, as I was shown a large hunk of meat in which tiny maggots could be seen writhing about, the epitome of revolting. For me, at that point in time, at least. Burmaa and her sister discussed this dancing meat in their language, but I got the gist of it: "It's full of maggots, we need to throw it out," Burmaa argued. "It's fine, we can just pick them out," her sister replied. This was done and, to my enthralled apprehension, the meat was cooked and served to me. Not only did it taste fine, but it was better than any of the meat we had previously eaten, more tender and flavorful, like a dry-aged steak. There are things our minds tell us are disgusting that are in fact highly delicious. We just have to disobey central command from time to time and live a little.

The Milk Foods of Mongolia

What follows is an incomplete look at the dairy foods that I observed being made from yak milk by a single family in one part of Mongolia. The nation has a large galaxy of dairy foods with regional variations; what I describe is simply one constellation. I have seen a diversity of approaches to cheesemaking within all the countries I have visited. This has shown me that most traditional cheeses are highly variable, with different communities constructing their methods to suit their needs. Techniques are introduced, modified, cast away, or revived. In short, tradition is not static, and it's probably rare that everyone in a given region makes the same cheese in the same way. Categorizing cheese and cuisine by country often misses the mark. Cheese, like life, rarely fits neatly within the borders we impose on it, and it often comes out better when we don't put up walls at all.

It appears that the boiling of milk is standard practice in Mongolia and that consuming raw milk is uncommon. There is good reason for this, as the bacteria that cause brucellosis are still found in the herds here. Brucellosis can cause extreme arthritis, weakening of heart valves, and central nervous system infections. It has been eradicated in most countries, as a simple test can detect the presence of the pathogenic bacteria that cause it, and infected animals can be culled. In Mongolia, huge herds are still moved across wide areas, and mandatory vaccinations and other forms of medical intervention are unevenly applied. A cultural response to the risk of zoonotic pathogens (those that can cross from animals to humans) spreading through milk is to heat-treat it by boiling. It is possible that the boiling of milk has been carried out in Mongolia and other regions not to eradicate problematic microbes but to alter milk's flavor and improve the texture of yogurt or the yield of some cheeses. I imagine there is no single explanation for why much of the world boils their milk, and I am not searching for a concrete reading of history's murky depths.

Immediately after milking, the yak milk was poured into a large, wok-like pot that sat over an opening on top of the woodstove. The milk was heated to a boil and allowed to simmer for thirty minutes, then poured into a large metal bowl. Notably, the same person who had just milked the animals did this, still wearing their milking clothes, without washing their hands or sterilizing the equipment. My eyes had been trained to look for possible vectors of what is called post-pasteurization contamination, and in this situation there were many. The milk was repeatedly scooped up and dropped back into the bowl, cooling it and causing an intentional foaming. Cream would rise into this foam overnight, and the low humidity caused a skin to form on top. In the morning, this cream-foam-skin was pulled off in a single deft motion by one practiced hand and dolloped into a saucer held in the other.

This cream-foam-skin is called *urüm* and can be compared to clotted cream and baked milk dishes. There are many names in many countries for this skin on the cream of boiled milk. I just refer to all of them as delicious. Urüm is eaten immediately and voraciously as breakfast, scooped and soaked up with little fried breads and washed down with mugs of milk tea. The thick cream has various textural layers: a sticky, almost chewable top skin, gelatinous cream goo on the bottom, and an airy, boiled milk noodle in between. It is eggy and expresses a pleasantly sulfurous minerality with a faint trace of the aroma of yaks. Yaky. I've found similar flavors in the milk of heritage cow breeds that feed on a grain-free, diverse pasture diet, so I'm not sure how much of what I tasted in this urüm was a true trait of yak milk. It is also richer than cow's milk tends to be, somewhere between cow and sheep in fat and protein content. Urüm can be churned into butter, although this seems to be becoming rare; cheap imported cooking oils are now common.

Removing the cream in the form of urüm was the first task of cheesing. The resulting partially skimmed milk was then made into a yogurt called *tarag*, which is a primary dairy food, source of sour whey, and basis for cheeses. The first step of making a thermophilic yogurt is to heat the milk above the temperature at which the proteins in milk denature, about 185°F (85°C). This was accomplished straightaway with the initial boiling of the milk. After urüm was pulled off, the remaining reduced-fat milk was heated until very warm, with a finger being used as a thermometer. Usually the pinkie, for some reason. I estimated the temperature to be about 120°F (50°C). The milk was cultured by adding a spoonful of a previous batch of yogurt; then it was poured into a metal milk can, which was wrapped in blankets to hold in the heat. After sitting overnight, it was thick and sour yogurt with a very pleasant lactic (sour cream) aroma.

Some batches of tarag were further processed into the most famous Mongolian cheese, *aaruul*. It belongs to an ancient family of dehydrated yogurt cheeses with variants being found throughout Mongolia, eastern China, the Himalaya, and across Central Asia to Lebanon and Turkey. It goes by many names, such as *kashk* in Iran and *qurut* in Afghanistan. The Mongolian version comes in a huge range of shapes and milk types, and often has sugar added to balance the acidity. Sweet-and-sour milk cookies is my analogy.

To make aaruul, tarag is boiled for many hours in the same large, wide pot that serves many functions in the ger kitchen and nests snugly on top of the woodstove. Small pieces of wood are used to get a hot flame going, with dried yak dung added in periodically to maintain a simmer. The yogurt gets thicker as it steams out its water, until it resembles pancake batter. This is poured into thin fabric sacks, which serve as draining bags, then hung to cool while a slow drip of yogurt whey is captured in a bucket. The sourness of the original yogurt has concentrated, so this whey is highly acidic and could be confused for lemon juice. It serves as the acid agent for coagulating other cheeses; I imagine it has a long list of further uses. Some cultures favor citrus or vinegar as their chosen culinary and cleaning acid. In Mongolia, the standard acid is fittingly provided by milk. This taking apart of milk into various components, aided by fermentation and heat and the utilization of so-called byproducts, is a hallmark of Mongolian dairy technology. As Burmaa told me, "We eat every part of the animal, and every part of its milk, because we love the animals and want to fully honor the lives they give, which we are required to take."

After releasing its clear whey overnight, the yogurt has become a paste with a texture similar to a fresh, high-moisture lactic cheese such as chèvre. Sugar is mixed in, then it is molded into shapes using pastry bags fitted with various tips, stamped into round cookies with customary patterns, or hand-rolled into long snakes or small balls. These are put on rectangular trays, which are placed on the roof of the ger to dry in the dependable summer sun, aided by frequent and prolonged steppe winds. They are turned and rearranged through the day and brought in at night. After two days, the aaruul is dry to the touch and can be stored indefinitely in cloth sacks. It doesn't get moldy left in this air, which desiccates the aaruul without turning it blue. Dehydration is the preferred approach to milk preservation in Mongolia, fitting in perfectly with both the climate and the lifestyle of herders. The removal of fat in the form of urüm serves to aid the dehydration and creation of a food suitable for long storage, similar to how parmesan and other hard grating cheeses are made with reduced-fat milk. Fat inhibits the expulsion and drainage of moisture. Lightweight aaruul can be stored in your pockets, carried in bags on a horse, or grated into soups. It is highly portable and resists damage during transport. It is sour milk with the water taken out, the perfect protein and mineral-dense food for the lives of mobile pastoralists.

The texture of aaruul ranges from a crumbly cookie to something so hard it will not yield to the effort of your teeth, causing your jaw to become sore in the attempt. In this form, it is sucked on and slowly dissolves, similar to a cheese called *chuurpi* made in the Himalaya. It can also be gritty or sandy, or break apart into shards and dust. The flavor is often a strong blast of sour milk, made less aggressive by the added sugar. Oxidized flavors are common, which I can only describe as being like stale milk, or butter left unwrapped in the fridge for too long. These come from the breakdown of milk fats or proteins during exposure to air, perhaps amplified by being left in direct sunlight. These oxidized flavors can be found in dairy foods all over the world, where they may be considered true to style and desirable to those raised eating them. I have yet to acquire a taste for these flavors, and I probably never will, knowing milk's potential. Not all tastes are worth acquiring, and there are limits to my omnivorousness. Until the food runs out.

The reduced-fat yak milk resulting from the removal of the cream as urüm can be used in a common beverage, milk tea. I would estimate the fat content of this milk to be 2 percent, based on mouthfeel and viscosity. After reboiling the unfermented skim milk with a small amount of budget Chinese brick tea, then adding many scoops of white sugar and some salt, this mauve-hued tea was kept warm in tall, insulated containers. It seemed like the family hardly drank cold or room-temperature water. They told me I would get sick if I did that in winter. When no tea was left, they would drink hot water. The tea was served piping hot in bowls or mugs, and the custom was to slurp it noisily. Although at first I found this slightly irritating, I adopted the custom, trying to mimic it in a flamboyant way, seeing if I could slurp even more loudly, to get a reaction. Outslurp the slurpers. The family simply smiled at me approvingly, as if I were becoming one of them.

These six months in Mongolia represented an unlearning and restructuring of my psyche. I had assumed that the way livestock were raised in the United States was simply how it was done. Seeing people in Mongolia successfully raise herds without fences or barns, while participating with livestock in an utterly different relationship of trust and understanding, caused me to examine my assumptions, to question my ingrained truths. Likewise, my thinking was shaken up by seeing that cheese could be made without all the purchased inputs, and with the same people who'd milked the animals making the cheese over a dung fire while not following any of the protocols I had been taught were necessary. I think of cheese as having four primary ingredients: milk, culture, rennet, and salt. As I learned in Mongolia, you can still produce shelf-stable, nutrient-dense cheese with only two of these: milk and culture. As will be revealed, perhaps all that is really needed is milk, and a vessel to put it in.

These older ways made sense and seemed to successfully result in tasty, healthy, safe food. I knew there were other refuges like Mongolia out there, where I could find these old ways, experience them briefly, and share what I saw with the outside world. I reluctantly left the ger camp, cresting the same pass by which I had entered, a piece of my heart breaking off and being left behind to sit under the sun and snow forever. I fervently hope that someday I will return to cross the pass again, back into the valley of the yaks, to hear herders sing the songs of the land as they ride proud horses across the sacred face of our mother, praising her with every step as they gallop out the rhythm of the earth's holy dance.

— CHAPTER TWO —

Goat Walking in the Italian Alps

ARPISSON, VALLE D'AOSTA, ITALY

> *The first of man's domestic animals to colonize the wilderness, the goat, is the last to abandon the deserts that man leaves behind him. For, ever the friend of the pioneer and the last survivor, the goat was never well-loved by farmers on fertile land. When agriculture produces crops that man, cow, and sheep can consume with more profit, the goat retreats to the mountain tops and the wilderness, rejected and despised—hated, too, as the emblem of anarchy.*
>
> —JIM CORBETT, *Goatwalking*

The concept of cheese trekking manifested in the act of doing it. The trail appeared as I walked it. Traveling like a snail with a backpack for a shell, visiting various parts of the world, I would bounce from one farm to the next, learning through participating in the work, mud on hands, shit on boots. Selling my labor for an hourly wage has never felt right. I prefer what is called a work trade: I agree to work in exchange for meals and a simple place to sleep. Beyond this, I can gain and share experiential, intellectual, social, and spiritual capital, which enriches myself and others in non-monetary ways. Beyond the specifics of making cheese, I was seeking to participate in the milking and herding of animals, to learn how farming fits into broader ecosystems, to immerse myself in the cuisine of a place, and to observe how people grow, collect, and cook their food. My heart told me to focus on experiencing as deeply as possible how people live, how their days are organized, how they communicate with their land and livestock, and what their relationships with their neighbors are like.

I sought to use language barriers as a tool, to go beyond the intellect to emotional and energetic levels, where we sense and communicate without words.

Instead of spending an extensive amount of time in any one place, I would bounce around to see the diversity of approaches and perform cross-cultural comparison. In hindsight, although the short duration of my stays limited my experiences, I feel this has been made up for by the number of countries, farms, and creameries I have visited. This has lent me a catalog of aromas, tastes, and cross-checks that have melded into what I now share with you: a deeper understanding of how a cheese can reflect the place where it is made, and what it may take to rebuild healthy and responsible dairy foodways again. These concepts need to arise from our senses and spread roots in our bodies before the intellectual fruits can ripen into something that can lead to lasting change. It all starts from the ground, and it's there that we will find the inspiration, the wisdom, the paths to understanding.

Andiamo!

I set out for a farm in the Alps above Valle D'Aosta, in northwestern Italy. Azienda Agricola Arpisson is an organic goat dairy with a separate, higher-elevation ***alpage***. Alpages are mountain farms where animals are brought in a seasonal **transhumance**. In the summer, a small revolving crew of volunteers come to Arpisson to learn by working in the stunningly beautiful mountains near the borders with France and Switzerland. After gaining elevation with relentless determination by train and bus, gunning up roads following cascading streams, I arrived in the touristy village of Cogne, perplexingly set in the bottom of a glacial valley. I was picked up by Paul, one of the volunteers, and we continued—up, up, and away from the world of traffic jams and busy plazas. As we drove the steep, serpentine road through hayfields, past large dairies and Swiss-looking houses, we stopped to chat with Attilio, the farm's patriarch, on his way down in his tiny, very Italian car. When I said, "I'm Trevor," he responded, "I know you are traveling, but now you are here to work." Indeed. A fiery, short man in his seventies who could likely beat me at wrestling, Attilio regarded me with sportive, green wizard eyes and the air of excessive confidence that some people feel the need to project. I liked him, and I was here to work.

The work was serious. The volunteers got thrown right into it and were allowed to sort out who was doing what with a level of autonomy. We woke at sunrise and after a quick breakfast of strong coffee and toast, some of us would start the morning milking of the herd of sixty Saanen goats. The kids were separated from the does and left in their romp room for the day where they jumped around and butted heads, acting as if they were grown-up. The goats rushed up a ramp into milking stanchions, and their teats were washed with wet rags and then dry ones, but no iodine was used pre- or post-milking. We didn't discard the initial squirt of milk that is often thrown out in America, a process referred to as stripping the teats. The logic of stripping is that the first squirt or two is especially high in bacteria and therefore potentially problematic or dangerous. What I would find in my travels is that in many countries people don't use iodine or discard the initial squirt of milk, yet the raw milk is still safe for cheesemaking. What I was seeing was a more relaxed approach to safe milk handling, one in which milk and microbes were trusted and food safety plans based on the farming practices, herd size, and the ability of the animals to spend the day in clean pasture rather than their own manure.

After milking, two of us would take the herd for a walk in the forested hills above the farm. One of the main things that attracted me to Arpisson was the practice of **grazing circuits**. I was interested in grazing systems like this that moved beyond fences and planned rotational programs. The goats would follow one person who walked fast in front, with calls of "*Andiamo!*" (Let's go), whistles, and a mimicked goat call that could be translated as "I'm over here." We tried to keep their momentum going as we climbed up past a network of trails popular with hikers and bikers. There were various clearings in the trees full of green understory plants; here the goats seemed to be happy, so we would settle down in these, letting them browse peacefully. Goats being what they are, we could only sit and rest for a few short minutes until some of them were wandering out of sight, always attempting to get the good stuff just ahead, then those in back would become jealous and move forward. The grass is always greener where the other goats are feeding. We circled around to push them into a tighter group, learning how to be a part of the herd, but also how to be high in the dominance hierarchies that are an inherent aspect of both goat and primate social dynamics. It was mushroom season, and I had to walk at the front to pick as many of the abundant chanterelles and porcinis as I could, because the goats relished them. I even picked some that had distinct goat teeth marks, where these epicurean nannies had taken a capricious nibble.

The modern, industrial style of dairying that is so popular in the United States involves bringing feed to animals. This feed may be grown in different states or continents. Besides hay and alfalfa, it often includes corn and soy as ingredients in grain supplements. The nutritional value of these feeds is scientifically determined, and they are mixed and fed to livestock to provide the high-energy input needed to maximize outputs of milk and meat. A whole other paradigm exists that, to my mind, is more sane, efficient, and healthy on many levels. This is the practice of taking the animals to their food, growing what is needed locally, or choosing how many of what type of livestock to raise based on what the land provides. In this other paradigm, animals are allowed to exercise their inherent nutritional wisdom, selecting the foods they need from a wide variety of vegetation to fulfill their individual needs and self-medicate by ingesting particular plants or minerals. These animals know what they are doing when they are allowed to be animals, when they can explore their surroundings, following the lead of a herd that shares a detailed spatial map, built from multiple generations of experience on a landscape.

Haymaking Is Cheesemaking

A major focus of Attilio's work was the putting up of hay. The practice of cultivating hayfields, which are then cut, dried, and stored to feed livestock in winter, is an established part of dairying, especially at high elevations and other places with a short growing season. Grasses and other plants transform abundant summer rains and long, sunny days into cellulose stalks and nutrient-dense leaves that cannot be processed by human digestion. Preserving this photosynthetic cycle and using it to feed livestock during the colder months is an integral part of sustainable dairying. We can generate our food (meat, milk) from the animals whose microbes can digest plants, and who provide the fertilizer to feed the pastures and hayfields as well as our vegetable gardens, orchards, and crops. Any proposal for sustainable food production that doesn't include livestock is missing a fundamental component of this cycle.

Proper haymaking requires gambling with the gods . . . and coming out on top. Once the fields are ready for their first cutting of the summer, all eyes are on the sky. Meteorology has always been intimately connected with prayer and prophecy. Rain gods are high-ranking figures in most pantheons. What is needed is a few consecutive days with no rainfall and warm, dry weather so you can cut the grass, dry it, and bring it in as hay before the next rainstorm. At this elevation, the first cut is in July, when dependable afternoon breezes aid the drying process. Only small areas are cut so that all the grass can be gathered in a few hours if dark clouds build on the peaks, threatening to crash the party. The grass is cut by tractor where possible, or with a walk-behind mower on steeper sections. The cut salad of pasture plants is raked into long, parallel windrows, which are turned over a few times the next day. On the third day, if the grass feels dry enough, a truck passes over the rows and scoops the hay up into a loose mass piled up in its tall bed.

The hay is unloaded, then pitchforked onto a conveyor belt, which drops it inside a gigantic hay barn. Slatted walls allow wind to pass through but can be closed during rains or once the winter snows begin in earnest. It's crucial that there is very little moisture, or the hay will start composting and can generate enough heat to combust and burn the barn down. One person inside would move the loose hay into the corners and pack it along the walls. The baling of hay into modern, square bales tied with twine and huge, round bales began in the 1850s and is a relatively new phenomenon. For most of history, hay has been stored loose, in conical haystacks or in lofts above where animals live to allow convenient feeding aided by gravity. Attilio had a sense of how to stack the hay from various fields in different sections so he could mix it to create a nutritionally consistent feed.

This awareness of how the feed is grown and processed offers a huge advantage. Attilio said he had moved out of the cheese room entirely but was still a cheesemaker. He made cheese by cutting hay, spreading the winter's manure onto the hayfields, and deciding when to move his cows to feed in a new area. It was beautiful to hear him talk like this, as this was part of the holistic view of cheesemaking that I was seeking. For Attilio, farming and cheesemaking are one thing, a continuum. The realization that I needed to extend my knowledge outside of the make room and into the barns and pastures was a major motivation for my travels. The only way to really know what is happening with the milk is to have an intimate connection with the herd, to be aware of their health, to be familiar with the pastures, and to know if there are any issues or abnormalities affecting a long list of variables that can impact milk and therefore cheese quality.

Tipo Fontina

The volunteers stayed at the lower farm where the goats lived, and a herd of cows spent the summer at the higher-elevation alpage. Two cheesemakers lived there and made all the tipo fontina cheese for the year. This is a washed-rind alpine cheese very similar to Fontina D'Aosta but made outside of the DOP-protected name. DOP stands for "Denominazione di Origine Protetta," the Italian designation for a food with a protected denomination of origin (PDO, a more general term) that has a certifying consortium and regulations. The consortium that governs the aging, distribution, and marketing of the DOP fontina is a big operation that deals in international markets. While the purported aim of protected name appellations is to protect traditional foods and their production methods, I was beginning to see holes in this approach.

Attilio gave me plenty of reasons why he has little respect for the DOP of fontina and other cheeses. They prescribe exact rules for how the cheeses look and are made but leave space for animals to be raised and fed in industrial ways. Before the DOP was established, many farms made fontina, but there were slight differences in the cheeses. Now it all goes into centralized aging and comes out uniform; any deviations are weeded out. It is said that this is in fact desired by "the market"; that "consumers" want predictable, consistent food. I'd argue that we are animals, not consumers, and that we didn't consent to becoming a market for globally traded commodities disguised as food. This whole model has been imposed and can be resisted. You can still walk through the valleys near Cogne and buy non-DOP fontina directly from small farms and taste slight variations and nuances that would be culled out of the DOP stuff. These single-farm cheeses are usually more interesting and probably truer to how the cheese was created in the past.

The DOP process undercuts the more traditional small-scale producers of fontina because the consortium can sell cheese at a lower price. Its export is subsidized by the EU, which also undercuts artisan producers in America, who may need to sell a similar cheese they make for twice the price. Perhaps forcing cheeses like those made at Arpisson to fit into the global market game is impossible to do while maintaining the traits that make them special and place-based. Maybe they should simply remain local products, sold direct at a premium price, through the avenues of tourism and local markets. What makes these cheeses vibrant is lost when they are made to be exported and forced to conform to the norms of industry.

Nothing beats tasting a cheese in the place it is made—walking in the fields where the animals graze, smelling their barns, their milking parlors, the spaces where milk is transformed into cheese. Walking half a mile, you might find another farm making a cheese with the same name—but it's a little different, reflecting a slightly different landscape that is maybe farmed differently, the fields sloping in another direction. It has its own herd, family, buildings, water, brine tanks, techniques. I know not everyone can do this, which is why I am here: to share the experiences I have had and reveal to you how powerful a voice a cheese can have. Most cheeses mumble, but others sing, scream, tell jokes, or speak solemnly.

At the upper alpage, the evening milk sits overnight in cans inside a basin with cold, flowing spring water keeping it cool while the cream rises. This cream is skimmed and collected over a few days until there is enough to make butter. During this time, it is stored in a cellar where it slowly sours. Every four or five days, the lightly soured cream is churned into butter in a small electric barrel churn. This butter was fantastic, bright yellow and bursting with flavors. I didn't get to eat much of it, as it was in high demand and so it was all sold.

In the past, there were rennetless cheeses (like *pultost*) made from the sour skim milk, but now a rennet-curd alpine cheese is made with the reduced fat milk. Cream is a concentration of fat rather than protein, so skimming reduces the amount of fat relative to protein. Different styles of cheese have their own ideal ratio of fat to protein. For a longer-aged, harder cheese, you want nearly equal amounts. Too much fat can impede the draining of a cheese and cause difficulties in creating a large, firm wheel that can be aged. Cow milk tends to have more fat than protein, which is why Parmigiano and many hard or long-aged cheeses are made with reduced-fat milk. By removing cream, you can change what type of cheese that milk is best suited for. This is done by feel by makers in many parts of the world. When the milk is richer, they take a bit more cream, to generate a cheese that can be aged longer.

Every day, the skimmed evening milk was put in a copper cauldron and whole morning milk added, warm from the teat. It tasted like the evening milk had been skimmed to 2 percent, and then, by adding whole milk, the total was brought to maybe 3 percent. The milk was slowly heated in the copper vat with a propane burner under it, without stirring. This made me nervous, as I was accustomed to stainless steel, which can scald milk left unstirred. But copper is the ultimate material for cheese vats, as it conducts heat better than any other metal, spreading the heat of the flame evenly across its surface, never getting excessively hot in one spot. Copper is preferred by chocolate makers for the same reason.

The treatment of the milk was gentle. There was no foam or broken fat. The milk was never pumped or stored in a refrigerated bulk tank. A packaged DVI starter was used, along with calf rennet. After coagulation, the coagulum was cut very small with a tool called a harp, made of wires strung like guitar strings on a metal rack attached to a long wooden handle. The curds and whey were stirred vigorously and cooked quickly up to 113°F (45°C). The curds were allowed to settle under the whey briefly, then removed in large masses to make four big wheels, which were pressed overnight before being salted in brine.

The cheese would sit for a few days to dry and begin aging in the cellar of the alpage. Once a month, the wheels would be hauled down to the lower farm, where there was a larger aging space: a true cellar, the *cantina* sat underneath the barn where the cows and goats spent the winter. Walking down the stairs, you passed from the warm summer into a consistently cool and moist layer of underworld air. A wooden door sealed in a dank, stone-walled milk crypt, with many wooden shelves holding well over a thousand pounds of cheese.

In the early days of aging, the young fontina wheels are rubbed with dry salt. This creates a slight layer of goop on the surface, as moisture is pulled out of the cheese and mixes with the cheese schmutz formed by the abrasive action of the salt. This helps kick-start what is called a smear rind, a sticky orange-pink biofilm dominated by salt-tolerant bacteria. After one week of this, the wheels are washed every couple of days in a salt brine made from spring water and 3 percent salt. The smear grows and releases a particular aroma that some would call stinky. The washing is eventually stopped, and the cheeses age out for three to six months, with common natural rind microbes growing as the salty smear layer dries up.

The upper alpage produced butter, aged tipo fontina, and ricotta—all from cow milk. Sour buttermilk was used as the acid to make ricotta from whey. In most cases, alpine cheeses have very little acid when they are removed from the vat. Because of the low acidity, the whey must have some acid added to get the ricotta to form. Adding acid seems to be more common with cow ricotta than with ricotta made from goat or sheep milk.

Caprino

At the upper alpage, things were rustic and time-worn, but the setup of the creamery at the lower farm was the opposite. The milk was kept cold in a fancy automated vat that doubled as a refrigerated bulk tank. Every two or three days, the milk was heated up to make a goat tomme referred to as a *caprino* (Italian for "little goat," used to signify a goat cheese). A tomme is a vaguely defined term referring to a small- to medium-sized wheel that is somewhere between soft and hard with a natural rind colored by molds. It looks more French than Italian, but this valley is very close to France, and the influence of that proximity is readily seen in the cuisine and dairy foods. In a modern riff, the culture for the raw milk goat tomme was kefir. A huge handful of kefir grains was kept in milk in the refrigerator, where it slowly created a thick sour milk. The strained kefir was poured directly into the 90°F (32°C) goat milk and fresh milk was poured onto the grains, then put back in the refrigerator until the next make.

I would not recommend fermenting your kefir this way if you desire an active starter for cheesemaking. There were too many grains that would ferment milk rapidly, so they compensated by keeping it in the refrigerator. It's hard to say if it was the kefir that was driving the fermentation or simply the bacteria indigenous to the raw milk itself. The most important part about using a natural starter is being certain it's at peak activity when you use it.

Liquid rennet from goat kids was added to the milk, causing a coagulum to form, which, at around forty-five minutes, was cut to the size of popcorn kernels. The curds were cooked and stirred until they felt firm, which tended to take a lot less heat and motion than with the curds from cow milk. Once deemed ready by the hand of the maker, they were removed into small plastic molds, forming wheels that were flipped several times before being rubbed with salt the next morning. This cheese aged in the cantina along with the tipo fontina. It grew a natural rind that saw little intervention, going through a succession of yeasts and molds. Spots of blue mold and fluffy gray mucor eventually gave way to bright-yellow and orange splotches, a rainbow of colors. All these molds are harmless and nearly universal, growing on cheese anywhere that the conditions are conducive. If the cheese ferments adequately, and enough salt is applied, the types of molds that appear are generally predictable and safe.

The whey left over was very milky, which is common with harder goat cheeses. The fragile curd is easily ruptured in the cut and stir, releasing the fat and protein that would otherwise be locked in the cheese. Many makers would see this as a "loss of yield," but the goodness is not lost. It can be retrieved through the magic of ricotta and is probably a big reason why whey cheeses are highly revered in so many cheesemaking traditions.

At Arpisson, neither fresh milk nor salt was added to the goat whey. The lack of milk and salt seemed abnormal to me, but I have now observed this in other regions of Italy, such as Abruzzo. You get much less ricotta without the added milk, and it has a more dense, less creamy texture. Nor was acid added to the goat ricotta; the curd simply appeared on its own, slowly, delicately. It was sold at farmers markets, and we consumed it as a staple food, especially at breakfast. I enjoyed it on toast with chestnut honey. After seeing ricotta made at dozens of creameries and making it myself with different milks, I am beginning to appreciate the no-salt, no-milk goat ricotta. It somehow feels like it doesn't need it. Perhaps because salt can mask its aromas, which, like everything about goat milk, are delicate. At Arpisson, the ricotta was dense with pleasant mineral and floral notes and a suggestion of the animal from which it came. As will be revealed, there is no one recipe for ricotta. The approach should vary with the milk you have, and with the cheese that the whey comes from.

The many lessons of Arpisson came to me through immersion in the physically demanding daily rituals of goat dairying. Walking the land with the herds as an active participant, I learned how livestock can be raised outside fences, according to a model of extensive grazing. This was the combination I was looking for; a blend of hiking with cheesemaking and foraging, a modern pastoral lifestyle that nourishes landscapes and animals including humans in kincentric circles. This demonstrated the ideal of procuring feed for a herd from the local landscape. It could be done! Here was an alternative to the industrialized paradigm that too many farmers have come to accept as normal and unavoidable.

There were deeper lessons emerging from the journey itself. They were not consciously in the front of my mind. My intellect was too busy processing the daily torrent of tangible skills and knowledge I was swimming in. The deeper, philosophical enrichment that was taking place was more subtle, happening in my core, rearranging furniture behind the scenes. Perhaps some teachings are best introduced on this pre-linguistic level, as felt or lived experiences that slowly coalesce in the heart-mind-spirit-soul complex. Later, once things slowed down and time for reflection became abundant, the simple, grounding wisdom of my time at Arpisson would produce fruit from the trees of this orchard that is not mine or yours.

— CHAPTER THREE —

The Cheese Is in the Milk

FINOCCHIO VERDE, PIEDMONT, ITALY

There are no separations between Jesus Christ, God and the Devil, because I am all of them.

—GG Allin

The wind horse knew what it was doing when it blew me toward the world's most important festival devoted to raw milk cheese, put on by Slow Food International every two years in northwestern Italy. It consists of four days of presentations, tastings, and seminars, but the real show is the narrow streets of the tiny town of Bra, packed with tables covered in cheeses, sometimes sampled out by the people who make them. This is the best opportunity I have found to taste a diversity of cheeses from small producers who work with natural starters. My mind, palate, and senses were permanently expanded, the doors of perception blown off their hinges. In Bra, I experienced a sudden awareness of what cheese could be. I realized that up to this point, I had only scratched the surface of the possible flavors in cheese—there was an unimagined depth below. Comparing the dozens of cheeses I tasted at Bra with what I had built up as a flavor memory bank in the United States, I understood how monotonous and uninspired most commercial-scale cheeses are, even artisan, raw milk, and farmstead ones. In their absence, I began to discern the generic flavors of cheeses made from commercial starter cultures applied to pasteurized milk.

I spoke with many makers while tasting my way through the streets in a fevered trance. One of them was an Austrian named Christian Lauer, who makes an old-school sheep cheese in the Langhe hills, just south of Bra. He invited me to come stay at the B&B he runs in a nearly abandoned village. He had come to Cascina del Finocchio Verde ("green fennel farmhouse") as a volunteer and stayed longer, learning how to make the cheese and run the farm, and eventually bought the place. On my way to Finocchio Verde,

I spent a night under my tarp tent on the banks of the Stura River, then hitchhiked into the hills, past monocultures of grapes grown for famous protected-name wines. Once-thriving farming communities that had possessed food sovereignty and grown diverse crops were now wine barrens, cultivating single varietals for export. Deep foodways are alive in Piedmont, but what we see now is a fraction of what once was.

Tuma

The Alta Langhe hills are the historical home of a beautiful little cheese referred to as a *tuma*, which in the past was made with the milk of the local, coarse-wooled Langhe sheep. Following the global trend, the local animals have steadily been replaced by modern breeds that produce more milk. Langhe sheep, often with tails left undocked and sporting wool-less legs, possess the hardiness and browsing (rather than grazing) abilities notable in many heritage sheep breeds. Grazing implies feeding on grass or plants growing close to the ground, whereas browsing means eating bushes, trees, vines, and other plants that grow higher. They eat acorns and woody brush, and thrive without most of the purchased inputs that "improved breeds" require. Despite its adaptation to the landscape, the breed had dwindled from forty-five thousand in the 1950s down to twenty-five hundred in the early 2000s.

The original tumas of the region likely had a level of diversity, with slight variations from one farm to the next, as each farm may have had slightly different feed sources, bedding materials, or milking procedures. Their creameries might have different ambient temperatures, and their techniques of handling the curd may have varied. Subtle differences such as these can be celebrated—or they can be ironed out to create cheeses that can be sold as a single entity, with a homogeneous identity, a consistent standardized product. The regional tumas have been subsumed by a DOP cheese called Murazzano, named for a village a short distance from Cascina del Finocchio Verde. The DOP regulations allow Murazzano to be made with an addition of up to 40 percent cow milk and don't specify which breed of sheep can be used. As a result, the connection with the breed of the region is lost, which is a crucial aspect of the identity of these tumas. Through this DOP, the *cheeses* of a place become a *cheese* of a marketing board. To address this, Slow Food created a presidium for what it called Langhe sheep tuma. The presidium calls for the milk of the Langhe sheep to be used exclusively if makers wish to participate. Slow Food presidia provide no legal status and have their own politics, but adhering to their guidelines can be used as a selling point by makers who choose not to join the DOP.

Rather than use the Slow Food or DOP descriptions, I would broadly describe Langhe sheep tuma as being in the *robiola* family of soft-ripened small wheels that are eaten from a few days old up to a few weeks, when they sport a light rind of yeasts and the early phases of white molds. It is a soft cheese, but not to the point of the paste breaking down into a gooey layer as seen in longer-ripened bloomy rinds like brie and camembert. It is dense and can be almost chalky in its core. The flavor is mildly bitter, barnyard and yogurt balanced with sweet sheep milk. The rind imparts mild renditions of the bread and fruit traits contributed by yeasts.

Finocchio Verde

I walked down a ridge above the wine barrens, into a crumbling collection of villas at the end of a seldom used dead-end road. It was uplifting to be walking, just me and my backpack, full of a youthful passion reminiscent of hitchhiking along the West Coast of the United States in my early twenties. Cascina del Finocchio Verde is a large stone house hanging off the end of a clustered collection of time-worn buildings. It alone appears to be permanently occupied, besides the tractors and hay that farmers store in the bottom story of once-bustling family homes full of laughter and heartache, now repurposed as cobweb-laced barns. Christian took a short break from his chores to show me to my simple room on the top floor, and we went to see a bit of the land. A tall, lanky, but strong and wiry young man, Christian moved with the deliberate certainty of a farmer with a long to-do list. We walked down a slope past a small vegetable garden that had been put to sleep for the winter, a few heads of cabbage and wilting fava stalks nodding off like the last ones left at a party.

After coffee the next morning, Christian invited me to gather the sheep for morning milking. We walked past low stone walls and outbuildings, then down dirt tracks with thick hedges on either side. The small flock of sheep were in a paddock made from electrified net fences, watching us approach. I bent down and put my fingers into the damp, black ground, inhaling the aroma of the living earth. It smelled of short days, earthworms, Halloween, and compost. Rotting leaves, soggy grass, mineral-rich soil, sheep wool, and their poop.

We walked back up the hill with the flock, under a gloomy, gray woolen-blanket sky. I would hesitate to say we were herding, because the sheep took the lead and knew the way home, heading there with a sense of purpose. Reaching the old wooden milking barn, they marched single-file up a short ramp and were milked out by hand while eating a small grain ration. Christian had been telling me about his cheesemaking process, and I was confused when he said he didn't add a starter culture. I kept asking what he meant, as it didn't square with how cheese worked in my mind, which had been molded in a certain paradigm and was only beginning to wake to other realities.

Christian began milking by hand, and as with the goats at Arpisson, he didn't strip the teats. When I asked why he didn't discard the initial burst of milk, he said, "That's the starter!" He pointed at the sheep's udders and proclaimed, "The culture for the cheese lives on the udder." It was an earthquake statement that altered the course of my river, shaking it into a new valley flowing to an unknown sea. It felt like I had been punched in the mind. "You can trust that the microbes native to milk will ferment it successfully into cheese?" I asked, incredulously. "Of course," he replied. "How do you think people made cheese for the hundreds of years before packaged starters were introduced?" I wanted to let my mind go there, but some part of me was holding back, insisting this was too sloppy and reckless to result in anything tasty.

He milked into a bucket, and when it was full, we carried it into the ***caseificio*** (creamery) and poured it into a large stainless pot. He makes the same cheese twice a day, after every milking, never storing the milk cold. We applied a touch of heat from a propane burner to bring the milk back to udder temperature. A French-made lamb rennet was stirred in before we took a break for breakfast. After the milk had been coagulating for two hours, a long knife was used to cut the coagulum vertically in both directions in a # pattern, forming columns of about one-eighth of an inch. We left it to rest for twenty minutes while we let the sheep out of the milking barn. They snacked on acorns, crunching them in their nutcracker jaws as they meandered slowly and insubordinately in the fog, on their way to a new section of pasture.

Returning to the cheese, Christian performed the cross-cut, where the pillars of curd we had cut vertically in a # pattern were cut horizontally into cubes. This was done with a shallow, sharp-edged disk that I have seen used in Italy and Austria. Tools matter. Their origin, feel, design, and history are a part of any craft. The curds were cut and scooped out in a single motion by sliding the disk horizontally through the vertical pillars of curd, severing them into imperfect cubes about the size of a hazelnut. There was no cooking or stirring to push moisture out of the curds, meaning the cheese would end up soft. The long coagulation time ensured that a lot of moisture was locked in, and that the soft but sturdy curds would not fall apart and release their whey.

Christian looked like a father placing his baby in a cradle as he gently placed the newborn curds into their beds: eleven small, circular molds with a single piece of cheesecloth draped inside them to let the wheels take shape. After the cheeses were flipped in their forms, the whey from the make was heated up and soaked into the cheesecloth, which was now used as a warm blanket covering the cheeses. A large wood box was preheated with steaming whey and placed over the cheeses on the draining table, keeping them cozy. Without the addition of a tangible starter, this heat was used to encourage fermentation in a cool room. The cheeses would acidify slowly, sitting for three days before being salted. At that point, they had begun to grow yeast in the humid creamery and smell like a beer-soaked sheep carrying a few bananas and salamis in a little pack on top of its earthy wool.

The cheeses were aged in an adjoining section of the creamery with no physical separation from the making space. An analog hygrometer (humidity reader) said the humidity was 95 percent, and Christian remarked that it was important to keep it above 90 percent to establish the proper rind consisting mainly of yeasts. Some makers have a dedicated space called a hastening room that is high in humidity and fairly warm to encourage yeasts to establish rinds on freshly salted cheeses. The entire creamery that Christian works in serves this role, and the cheese never has to be moved to a cooler, cellar-like space to age further. I'd never seen someone make cheese in the same room where it ages. It struck me that integrating these functions in a single space must set up a microbial feedback loop where the spores of the fungal rind can establish more easily, as they are in the air and all over the equipment, falling in the milk as it is made into cheese. By making cheese twice a day, Christian is feeding a steady supply of cheese to the resident SCOBY of his creamery. It was a clean space, kept rigorously tidy, that smelled like angelic cheeses who were tempted by the dark side, whispering earthy secrets of fungal forest floors and hay starting to grow moldy in a damp cellar.

After seven to ten days, a light, white dusting started to appear, a fungal community including a form of ***Geotrichum candidum***. Referred to as *Geo*, this common fungus that loves living on cheese is a yeast (single-celled fungus) that can become a mold (multicellular, filamentous fungi) as it matures, with some strains being more likely to become visible molds. Microbiologist Benjamin Wolfe refers to it as "a yeast with moldy tendencies." *Geo* can lead to a wrinkly, brainy rind when its filaments spread and stitch into a weblike mycelium on the surface of the cheese. It consumes lactic acid there, and contributes earthy mushroom flavors while discouraging other microbes from establishing in the early succession stages of the rind.

The *Geo* and other yeasts growing on this cheese do not originate from a package. They could be a part of the milk microbiota, from the farm environment and pastures, or residents in the creamery, released into the air as spores by the mature cheeses. Maybe they come from all these sources, or from somewhere else. They didn't originate from a laboratory in Denmark—that we can say with near certainty.

Satori

I put my face up close to one of the cheeses nearing its peak age of ten to fourteen days. I breathed in deeply and was taken aback by what infiltrated my senses. The clock stopped ticking. The cheese smelled remarkably like the field we had walked in as we approached the sheep before morning milking. Living soil, rotting leaves, wet grass, the lanolin of sheep wool, and their poop. It was a stunningly pleasant and tangled knot of aromas, complemented by hints of mushroom and cellar from the sprouting white mold, all tied together by the fruity esters of yeasts.

That the smell of this unassuming little cheese with its microbial wilderness could induce this state of rapture in me as it released its spores and aromatic compounds into the air was revelatory. It was my first cheese satori, to adopt the Zen Buddhist term referring to a sudden enlightenment, a feeling of total presence, an extreme, consciousness-expanding epiphany. That a cheese could have such a characteristic personality made real what had been just a concept before, introduced to me in that Seattle bar ten years earlier. The notion of terroir crept out of my brain like a spreading fungal mycelium and moved through my whole body. Some cheeses are consciousness altering, more entheogen than food. These few moments of heaven are locked permanently in my flavor memory bank, setting a precedent from which I would spend the next three years recovering. Chasing that first high, I would venture into the forgotten folds of the planet where other boundary-dissolving, cheese-born intoxicants lurked, guarded by faithful initiates, persecuted by the servants of an empire that fears everything unabashedly sexual and of the earth.

I had never experienced such an undeniable sensory experience of what I would later dub **deep terroir**, where a direct aromatic correlation exists between a cheese and the place it is made. That place wasn't just a creamery, mind you. It was a field of plants about to enter winter dormancy, growing from wet, fecund September earth on a west-facing hillside above a tributary of the Po River in the Alta Langhe hills on the piece of Planet Earth that is temporarily called Italy. The creamery, with its pleasant aroma of souring sheep milk and the yeasty rind of the tuma, could be viewed as a carrier of the terroir seeded by the land and animals. An amplifier, the space in which the seeds of this cheese germinate. That field and this cheese smelled of sheep, with hints of their manure as a seasoning in this aromatic stew. Not in a bad way, but in a pleasant, complementary way. It wasn't the aroma of many sheep in a closed barn, packing urine and manure into an ammoniated cake without enough interwoven dry plant matter. The aroma of that field included as one element the input of a small flock of healthy sheep with plenty of space, being moved daily. It was a light sprinkle of ovine biology that didn't overwhelm the other flavors in the stew—the wet grass, dark soil, and rotting leaves. The sheep lay in that field, breathing its air, eating its plants, and walking its contours. The milk carried a certain essence that I couldn't identify when drinking it. It was likely very subtle, below the threshold of perception for most of us. This essence was then somehow magnified to a perceivable level and altered by the fermentations involved in the cheese make and the subsequent short ripening. Milk can have hints of terroir, but it is really the process of turning it into cheese that can unlock this potential and allow a coherent voice to start speaking.

This cheese is one of the greatest I have tasted, based on that initial powerful sensory imprint left by smelling and then eating it in the place it was produced. Terroir in cheese went from something conceptual to a highly tangible, blissful experience. There was no packaged starter added—the cheese slowly fermented as the bacteria native to the milk sprouted in the infant cheese, into a heavenly cloud of moldy milk that brought me to the verge of a mystical realization. There are cheeses you eat, and there are cheeses that let you shake hands and sit down for a drink with God.

A Cheese from Hell

Paradoxically, the divine tuma can also be made into one of the most off-putting, hellish cheeses I have ever bumped into: *bruss*. This is of course a highly subjective call, and for the record I travel far and wide to taste cheeses that I expect to find unpleasant. I am seeking to understand how cheeses fit into cuisines, local farming models, and the lives of people. I want to learn to like, appreciate, or at least understand foods I don't enjoy. When I make pejorative statements about a cheese, I don't mean to insult it, its producers, or the cultures to which it is attached. Our experience of eating any food is always culturally and psychologically shaped and filtered through layers of memory and association. While acknowledging all this, I still feel it is valuable to state in sometimes blunt terms how I react to a cheese, to begin with my immediate sensory experience: *Bruss is made by Satan, with the fiery terroir of the sulfurous pits of hell.*

The tuma has a short one- to two-week window of peak ripeness. At times some wheels ripen past this and begin to dry out or perhaps get overly funky, covered in colorful molds. To address this eventuality, a practice evolved of cutting the past-prime cheeses into chunks and mixing them with milk to form a paste. This usually includes grappa, an alcohol distillate made from the grape debris remaining after juice is pressed out for winemaking. The cheese-grappa mixture is packed into jars, sealed, and allowed to age anaerobically (without exposure to oxygen), sometimes for months. This is to ensure that cheesemakers will have something to eat during the winters when they are not milking. Old-timers are said to be the biggest fan base of bruss, a recurring story surrounding strongly flavored cheeses born from resourcefulness.

Christian grinned mischievously as he served me bruss on toast at the end of one of his farmhouse dinners. In retrospect, I think he likes feeding this to unsuspecting outsiders, like a teenage delinquent getting younger kids to take their first puffs of a cigarette, then laughing when they cough. It is beyond what is often termed *picante* in Italian, which can mean "strong" or "sharp" in reference to cheese. The word is related to the French term *piquant*, literally translated as "pricking" or "stinging," which often means pleasantly sharp or stimulating regarding food. It tingled my mouth with the numbing sensation induced by Sichuan peppers, combined with the nasal burn of wasabi. I chased this acrid lactic monstrosity with red wine as instructed, which tasted like muddy water under the influence of the infernal *formaggio*. I soon went to bed, mildly annoyed at having ended such a good meal and day of cheese revelations with a gastronomic horse kick to the face. Waking up the next morning, I could still taste it. To this day, I can still taste it, the evil demon twin of an angelic saint of a cheese.

— CHAPTER FOUR —

A Land Where Milk Grows

INNLANDET AND SOGNEFJORD, NORWAY

We live at the edge of the miraculous.

—Henry Miller

When friends from a cheese organization called Norsk Gardsost invited me to spend six weeks visiting cheesemakers in Norway, I jumped at the opportunity. My main interest was to experience a regional agricultural practice known as *seter* farming. Seters are mountain farms similar to alpages that feature small wooden homes and outbuildings set among pastures and woodlands. They are only occupied during summer months in a vertical transhumance in which livestock from permanent lower-elevation farms are moved up once snow has receded, giving them access to the greenery that explodes as the short summer gets to work. This allows the pastures around the home farm to rest and hay to be grown, along with crops of grain, corn, or vegetables. This type of transhumance is one of the main forms of **agropastoralism**. Only 3 percent of the country is considered arable, so any land conducive to growing crops is used for this purpose. Livestock are happy to utilize the steeper areas, where they put on weight and get exercise while enjoying the clean air, cold water, long days, and rich diet of the bountiful northern summer. It is said that humans do the same.

Globally, the act of grazing animals in mountainous areas often creates and maintains the open parklike meadows found in subalpine settings. This is one example of how livestock can be an integral part of a healthy, functioning ecosystem, and how human-livestock symbiosis can increase biodiversity. Grazing can expand the number of edge zones, known as ecotones, where different plant communities meet. In the transitional space between forest and meadow, a unique habitat is created, home to an edge community rich in flora and fauna. These in-between zones with fuzzy edges occur naturally, in riparian plant communities along the banks of a stream, or along the edge of a landslide or an area burnt by wildfire. Sometimes a physical boundary created by humans and composed of organic materials is part of these biotic overlaps, such as the hedgerows of the United Kingdom, or the agave and cacti living fences seen in Mexico. Stone walls can create habitat for rodents and insects, plants, and fungi. Ecotones are noted for their ecological diversity, often displaying a larger number of species than is found on either side of the boundary. The edge is where the action is.

Viewed from above, modern farmlands resemble a grid or checkerboard of different colors with pronounced divisions of land usages along right angles. These may cover many square miles of relatively flat, arable, watered land: the areas that are most sought by agriculturalists. In the western United States, you often see not a grid but giant circles, resulting from center-pivot irrigation. The bird's-eye view shows a stark contrast between the irrigated, green, cultivated zones and the surrounding tan or brown areas that receive little rainfall. The realm of human control and domestication is on one side. This is the side of civilization, godly order, progress. On the other side are the nonhuman, uncivilized, messy, untamed, devilish, wild places to be feared or subdued with a sharp, arbitrary line. A line violently defended. Cross the border and we'll kill you, whether you are a plant, dog, microbe, or human.

In other parts of the world, things look less distinct, a bit more chaotic. In many traditional farming systems, the blending of various land uses results in a mosaic of habitats. This creates landscapes that resemble patchwork quilts of irregular shapes, not always conforming to straight lines and right angles or perfect circles. Waterways meander, terraces traverse contours, property lines follow watersheds and ridgelines. These more hodgepodge, organic formations offer a visible representation of how human agriculture, settlement patterns, and foodways can be immersed in functioning ecosystems. In these situations, you find lots of edges, blurred boundaries, an increased diversity of landscapes.

These mosaics include uncultivated areas of woodlands, swamps, scrubland, or desert. Such zones may be used by humans for certain purposes, or left essentially free from human interaction, what many would call wilderness. This term represents a Western perspective that sees land as either part of the human sphere or not. This dichotomy underlies the conservation mentality that sees wildlands as places that we should only visit temporarily, leaving no trace. The implication is that human impact is inherently negative.

This current runs deep in the Western psyche—the feeling that we are not quite at home on this planet, that we are inherently predisposed to damage ecosystems. That we are separate from the land on which we live. That nature resides in the places we have left alone, the places where we are not.

Seters are often located amid forests or near tree lines with more open, alpine areas above. Wooden fences are constructed from local timber. The homes and outbuildings are generally surrounded by heavily impacted areas sometimes referred to as sacrifice zones, as it is difficult to avoid overgrazing when livestock cross these areas frequently, loafing and roughhousing there. As you move out from these mountain farms, there are concentric rings of less impacted forest, meadow, or shrublands, often with coppice groves and woodlots found in the areas close to the central buildings. The animals carve trails that radiate out through the rings, into less impacted areas and down to water sources.

These farms are examples of lands that are neither wild nor completely domesticated. They are somewhere in between. The buildings themselves reflect this, being made from local timber, topped with green roofs of thick, living sod. These ingenious roofs are highly insulative, keeping homes cool in summer's heat and holding in the warmth from a woodstove in cold weather. The root systems of plants spread and knit together individual pieces of sod, re-creating an organic, living whole. The houses blend into their landscape, a visual representation of living *in*, rather than on or off, the land.

The Butter-Cream Path

The older styles of Norwegian cheesemaking start with the isolation of milk's higher self: cream. And cream grows into its further purification: butter. For Norway, this was historically the most important dairy-based trade item until the twentieth century, as butter was valued as a cooking fat and served as a form of currency. Just as charcoal is a more efficient trade item than firewood due to its higher heat potential in relation to weight, butter is a denser concentration of calories than milk or cheese. The origins of butter making are impossible to determine, as butter can begin to form through the inadvertent churning of milk, cream, or yogurt during transportation. When agitated enough, milk fat globules rupture, and fat is released as an oily yellow substance that can be removed by hand. This likely happened in many places where milking was adopted and techniques for replicating this delicious accident were developed and spread through cultural diffusion. In some places, churning is still done inside an animal skin or wooden container with a plunger.

There are multiple approaches to separating cream from milk. The most common method today is to remove the cream by putting it through a centrifuge that can be electric or hand-powered. The older methods involve leaving milk overnight in a cheese vat or wide, shallow dish, then skimming the cream by hand with a wide ladle or releasing the skim milk from a hole in the bottom, leaving the cream behind. Another way to make butter is to skip the removal of cream and make whole-milk yogurt, then churn that. This works especially well for non-bovine milk, which does not form as distinct a cream line due to its smaller milk fat globules.

It seems that, historically, separation would have happened spontaneously, as milk was left in wooden barrels or troughs where it would sour as the cream rose, transforming into sour cream sitting on top of a gel of delicately coagulated skim milk, facilitating an easier and more thorough removal of the cream than pulling it off liquid milk. The lactic acid bacteria necessary to ferment the milk and turn it sour are already there, in the milk. You can leave healthy, fresh raw milk in a jar, and it will eventually sour and coagulate. This is known as a **spontaneous fermentation**, since nothing tangible is added to seed the milk with bacteria. Once this first round has soured, a small bit of it can be added to fresh milk, a process that many call backslopping but which I refer to as carrying forward. Repeating this process results in a sour milk culture called clabber, which can be drunk like kefir or used as a cheese starter. In the case of milk stored in barrels, wood can hold on to a residue of the culture, carrying it forward. The *rømme* (sour cream) thus obtained would be churned into butter in a hand-turned, coopered wood barrel or a tall cylinder with a plunger. Churning results in butter and buttermilk. Buttermilk is to butter what whey is to cheese: the liquid remaining after the solid has been removed. Buttermilk is cream with the butter taken out.

In the past, rømme was rarely consumed as sour cream since it was made into butter, a critical income stream for the farm. Now rømme is highly prized and abundant on the seters, where it is popular with hikers and tour groups, who enjoy it on-site, served on pancakes with jam or fresh berries and sugar. I was told by more than one Norwegian that it would be a shame to churn the rømme, which yields a much smaller amount of butter. Sour cream is such a tasty and endearing emblem of seter farming. Like the brief summer grazing season, it is ephemeral and blooming with exuberant life. When you have a food this perfect, why wait or attempt to go any further with it?

Historically, sweet butter was probably much like sweet milk—rare or nonexistent. What we call cultured butter was the norm. Butter was "cultured" because ferment is what milk does. By simply allowing microbiologically rich milk to sit in a vessel that could never be sanitized, souring occurred. This fermentation made the cream and resulting butter easier to obtain, tastier, and more shelf-stable. Many modern butter makers have found that letting cream sit in a cold space for two to three days improves the yield and sensory quality of butter. This is part of the revival and reintegration of what still are sound, safe, and delicious approaches to storing food and living in alliances with microbes.

I thought I knew what sour cream was before I arrived in Norway. My knowledge was shown to be very shallow by my first experiences with rømme. Most make it by separating cream from sweet milk, then culturing and allowing a light fermentation to take place. This no-longer-liquid but not-quite-solid sour cream expresses the delicate floral aromas of the biodiversity surrounding the farms. At its best, it has a subtle acidity but is still somewhat sweet. It coats your mouth in a transitory sheen of fat that multiplies the appeal of just about anything it is served with or on. Potatoes, fish, fresh berries, bowls of porridge, and roasted meats, for example.

Braskerudsetra

I first experienced real rømme at a seter called Braskerudsetra. I was welcomed warmly by my hosts Eric and Roy on an evening in July, the sun still shining at nine p.m. They are stout, thick-armed farming men with trimmed and combed beards, and gentle demeanors that match the quaintness of their picture-perfect mountain farm. We sat down for a late dinner that included green salad, potato flatbread, butter, and cheese. What stood out was a mound of little boiled yellow potatoes and a large bowl full of rømme, which I repeatedly took spoonfuls of until my squashed potatoes were doing lazy backstrokes as I rained coarse salt onto their giggling, golden, still-steaming bodies.

The more I mashed the potatoes into the creamy substrate seasoned with salt and pepper, the more I appreciated the ecstasy of the pure expression of this place and its culture through milk, concentrated in cream, and fermented into terroir by microbes. Many layers deepened the experience of this sour cream: the old cattle breed, the diverse diet and pure mountain water the cows were raised on, their health and fitness based on the knowledge and attention of the farmers, the generationally transmitted technical knowledge of how to ferment the cream to just the right point. It was a perfect meal, shared with folks I had just met who had assembled the essence of seter farming onto a single, well-used table.

A meal is more than the food on the plate. It's the combined psychology and overlapping energetic fields of all the people at the table, the plants and animals being consumed. It is the space in which the table lies, and the wider setting with its acoustics, memories, spirits, and subtle forces. A meal like this *is* the place. The food is part of a web that extends outward, off the table, through the kitchen, out the windows, and into the barns, fields, streams, moss-covered stone walls, and dark forests. A mycelium threaded up off the plate and into my stomach, my heart, then out my body and back into the rind that is the earth's crust, that holds the planet together.

The deliciousness of that meal and those creamy potatoes also conjured memories of my grandparents. My grandfather Carl was the son of immigrants from Norway. He married my grandmother Irene (who was of German descent) in a small agricultural community in rural North Dakota, where they farmed. Carl was proud of and identified strongly with his Norwegian roots, even after he lost his farm and moved to the Puget Sound in Washington State during the Great Depression, along with many Scandinavian and German economic migrants from the middle states. I have a particular aroma etched in my memory: the smell of my grandparents' kitchen and dinner table. I could try to dissect it into relatable particulars—such as pancakes, coffee, roast meat, boiled potatoes and cabbage, freshly baked bread, cinnamon, pickled fish—but then I would be tearing apart what is a unified aroma and lose the magic in doing so. The more immediate associations I have are emotional: comfort, belonging, loving care, an aura of warmth and safety around a table of laughing family while it is drizzling, dark, and chilly outside.

The emotional aspects of this aromatic memento were evoked powerfully and immediately upon entering the wood cabin at Braskerudsetra, although I couldn't attach them to a place and memory at first. Without my awareness, the impact of that meal was filtered through these subterranean memories, which added their emotional color and texture to what I perceived. Our experiences of eating are filtered through many layers of subjectivity: memories both conscious and unconscious, culturally influenced judgments, and personal history. Our sense of smell can evoke such intense emotions as it re-creates the bodily sensations of moments from our past that I consider these experiences to be a form of time travel. Our conscious minds may not be able to place certain memories on a timeline, but the body never forgets.

My first morning at Braskerudsetra, I got to observe some of the work that went into the seemingly simple meal. A herd of red-and-white cows were lined up, impatiently waiting to be let into the milking barn. These regal ladies had horns tipped with little gold balls to prevent them from gouging one another. They marched into the barn with purpose, each cow going to a space they were assigned not by the farmer, but by the herd's own social structure. The passing of many winters had worn the life-encrusted walls of the dark, wooden cow shed into an artistic expression of elegant perseverance. It smelled pleasantly of the bodies and manure of cows, of earth, plants, and compost. Beams of light penetrated, dimly lighting the scene as Eric and Roy made their rounds, attaching teat cups and hoses to each animal, causing milk to flow into an overhead pipeline. The mechanical drone of these pumps and rhythmic clicking of a pulsator make up the auditory cue that milking is happening, and can be heard outside millions of barns, homes, and sheds around the world.

On some days the milk goes directly into a bulk tank, to wait for a truck that picks up for Norway's milk company, TINE, to be lost in an anonymous field of milk that is plowed (pasteurized) and made into commodity foods. This is important income for dairy farmers around the country, but it strikes me as sad to see this glorious, place-based, microbially rich mountain milk being stripped of its provenance, made uniform by a corporate food system that fears, rather than reveres, life. On other days the milk is diverted during milking straight into the electric centrifuge separator, while it is still udder-warm. The cream is cultured with a common commercial starter and left to slowly ferment in a cool room or refrigerator until it is deemed ready to be churned into butter or packaged as rømme.

Pultost

Wherever cream is removed and butter is made, the question is always: What to do with the skim milk? It is this milk that leads to an obscure family of rennetless sour skim milk cheeses that were made in many butter-producing regions before the introduction of the rennet-curd approach. They have a particular texture due to being high in protein and low in fat. Some of them exhibit a peculiar crumble that tends to break down into a gummy, translucent gel, sometimes as a halo around the outside of a large wheel. As the rennet-coagulated cheese styles common today became more popular, this ancient family became marginalized and now only exists in refugia, small corners of the globe where they are often kept alive as cultural emblems and symbols of resistance to an industrializing food system. Norway is still home to a few members of this alienated, nearly forgotten family. *Gamalost* gets a lot of press as "the Viking cheese," but its weird little cousins *pultost* and *skjørost* caught my interest.

The process of making pultost and skjørost involves a pure **lactic coagulation** with no rennet, and a slow application of heat to firm up the highly acidic curds and whey. At Braskerudsetra and other farms I visited, the cream is skimmed during milking, with the warm milk going through a centrifuge that splits it into two streams: thick, yellowish cream from one spout; thin, white skim from the other. The skim milk is cultured and left in large plastic bins to sour and coagulate into clabber. As a week's worth of the clabbered skim milk is poured into a two-hundred-year-old handmade copper pot, the coagulum is broken into curds of various sizes. This is the equivalent of cutting the curd, separating the unified coagulum into curds and whey. The oldest bins have rafts of yeast on top and the early stages of a wrinkly biofilm that likely includes *Geo*. The heat is turned on very low and the curds and whey cooked without stirring for about three hours. The delicate curds float on top of the whey as the temp slowly rises above 104°F (40°C). The curds have firmed up significantly at this point, and some are removed to become skjørost: a fresh, squeaky, low-fat cheese, like cottage cheese curds without cream. These curds are ladled into a cheesecloth with plenty of hot whey and hung to drain for one hour before being crumbled up, packaged, and stored cold. Skjørost is used as table cheese, put on potatoes, and mixed in with rømme to create something similar to cottage cheese. The textural contrast is quite pleasant, one of the many ways to entertain the senses by taking milk apart then reassembling it in different configurations. The textures as well as the flavors that can be unlocked from milk are infinite.

Most of the curd is left in the vat and heated for another one or two hours, up to about 145°F (62°C). The curd raft bulges and is flipped over, then hot whey from underneath is poured over it to make sure the heat is evenly distributed. A test is done by squeezing a handful: If the curd comes together into a solid, lumpy dumpling that can be tossed and caught, it is ready. It has a mouthfeel like a very dry, slightly gummy ricotta. The curd is hung in cheesecloth as a clear whey drains out, then the large balls of cheese, still coated in cloth, are stacked in a wood press and pressure is applied to remove even more whey. That night, the pressed cheese masses are crumbled up into pieces the size of corn kernels and rice grains, inside a wooden trough. They are left in a mound in this vessel that is carved from a single tree trunk and likely serves as a reservoir for the fungal community that carries out the next phase: a secondary fermentation necessary to transform the unassuming fresh curd into pultost.

This mounding begins a strange process I had never heard of before. During exothermic fermentation, the mound begins to produce heat much the way a compost heap does. The small size of the broken-up curds maximizes surface area and airflow while mounding holds heat in. Apparently, all cheeses release some heat as they ripen; pultost just takes the process to an extreme. Yeasts grow rapidly, throwing off this heat and making the cheese less sour as they consume lactic acid. This sets off a chain of metabolic reactions that lead to the specific textures and flavors of this endangered, esoteric cheese.

By the next evening, the mound was releasing strong aromas of acetone and bananas. "That means it's ready," Roy told me, smelling the air as we walked into the creamery. Placing my hand into the curd, I was shocked that it was almost hot. The pile of cheese felt alive. If I had left my hand in there long enough, it might have been stripped of its flesh. We opened the windows and broke the curds up again, spreading them flat in the trough this time to encourage the heat to dissipate. The curds stay in the trough for multiple days and take on a distinctive warm-beach-meets-cow-barn aroma. The texture softens to a chewy gumminess and the color goes from white to yellow, with a transparency that is a hallmark of ripened sour skim milk cheeses.

After about one week, the pultost is finally salted and a hearty dose of caraway seeds added. It can be eaten at this point or left to ripen further for three to four weeks. In some instances, the curds melt down to a porridge-like puddle that can be spread on flatbread or potatoes. It's a strange cheese, and I wouldn't say I ever developed a taste for it. It is fascinating and noteworthy for what it represents: a philosophy of using every part of milk to make regionally specific foods—not for commercial purposes, but to feed people who feed livestock who feed life. Here is a cheese that remains intimately rooted in the place it is made, entangled with the cycles of life and nature from which it emerges. Cycles that include humans as active participants, rather than as outsiders separated from life by the illusion of discrete boundaries.

Brunost

The word *brunost* simply means "brown cheese" in Norwegian. It is really a family, with many branches and hybrid styles. Where pultost is a result of the extreme souring of skim milk, brunost goes in the opposite direction. Sugars are concentrated rather than fermented, into something more confection than cheese. Pultost is obscure and endangered, while brunost is Norway's most famous cheese and in no danger of going extinct. It is made on all scales of production, from small farms making twenty-five gallon batches over wood fires to the national dairy cooperative TINE making shrink-wrapped blocks seen at every grocery store. While brunost may seem highly "traditional," it represents a shift away from the original butter-cream path to the more recent rennet-curd traditions. The difference between radically new and traditional is about fifty to sixty years, an arbitrary distinction. Tradition is a creation of the present that's transposed on the past, a slippery, oversimplified concept.

Brunost is made after the production of a rennet-curd cheese, as a means of extracting residual nutrients from the whey. Whey is most of the liquid portion of milk that contains the constituents that didn't get trapped in the three-dimensional lattice of the coagulum. Roughly 80 percent of the proteins in milk are known as caseins, which allow coagulation to occur, and ideally all of the caseins get incorporated into the coagulum. The other 20 percent are known as whey proteins, meaning they evade coagulation and end up in the whey. This includes albumins, a protein category also found in eggs that can give ricotta a faintly eggy character. Whey also contains a small amount of fat and minerals and, most important, sugars in the form of lactose.

It is this lactose-rich whey that is prized for brunost making. Care is taken to harvest the whey while it is still sweet, before fermentation has progressed. First a rennet curd is made, a category referred to as white cheese in Norway.

After this curd is removed, the whey is heated to kill off the sugar-fermenting LAB, introduced by a starter or indigenous to the milk itself. This halts the fermentation and prevents digestion of lactose into acid. The aim is to extract these sugars, which is accomplished by boiling the water out of the whey. Unlike rennet coagulation or the heat-and-acid combination used in ricotta-style whey cheeses, brunost involves a concentration through dehydration similar to making maple syrup, where tree sap is reduced to a fraction of its original volume, concentrating its sugars and making it shelf-stable.

After most of the liquid has been boiled off, the thick liquid takes on a golden-brown hue. Finished brunost is often compared to caramel, and many believe it is caramelization of sugars that is taking place. In fact, it is mainly prolonged **Maillard reactions** happening, as sugars react with amino acids. The proteins in the liquid break down during the first hours of cooking. The building blocks of these proteins are amino acids, which are now freed and begin interacting with sugars to create the elevated sweetness and umami richness that characterizes brunost. Similar processes happen when grilling a piece of meat or browning butter—these are referred to as Maillard reactions.

To learn about brunost, I visited a small goat farm called Skjerdal Stølsysteri, which sits above picturesque Sognefjord, Norway's largest fjord. From the lower farm, guests walk up to a mid-mountain café built onto the creamery and a goat barn where they can enjoy a cheese-focused lunch while taking in the view. There is no bulk tank to store milk cold, so the milk is pumped into a vat and made into rennet-curd wheels twice a day, shaped in antique wooden forms with crosses on the bottom. The wheels drain over a copper-bottomed vat that heats the whey up to a boil over a wood fire, stirred by a large, purpose-built machine that resembles a giant dough mixer. The whey reduces for many hours to a thick brown syrup before being set aside to cool. Every two to three days, multiple batches of this reduced whey are combined and finished over high heat with constant stirring. The whey becomes a thick brown paste as steam billows out, mingling with the smoke from the fire that burns from dawn to dusk. When deemed finished (a subjective call made by sensory observation), the vat is swung away from the fire and stirred even more vigorously as it cools, to prevent excessive crystallization, which can lend an undesirable sandy texture.

After cooling to the point that it can be handled, the golden mass has a texture like sticky modeling clay. It is flung onto a table and pressed firmly into antique wooden molds, forming rectangular pillars often decorated with ornate Nordic patterns. At Skjerdal, it is sold on-site to guests who are served a whole pillar, along with the Norwegian national beverage: black coffee, which I found to be more common than aquavit. Coffee pairs well with many mild hard cheeses, and is a perfect match for caramelly brunost. Folks take turns pulling slices off the block and enjoying it on top of pancakes with strawberry jam, a classic pairing. The tool used to obtain the paper-thin slices is called an *ostehøvel* (cheese shovel). Every time I was served brunost in Norway, it was accompanied with this tool. Some cheeses are so cool, they have their own knife.

Both pultost and brunost involve heating milk to high temperatures, which affects the milk microbiota in a way comparable to pasteurization. I would argue they still possess a taste of place. Terroir in cheese is not only driven by the plants that animals eat or the microbes in the milk. Humans are a crucial component, and it is this cultural level of terroir that is often missing from the discussion of what gives a cheese a sense of identity, of being from somewhere, of representing a place. It is through human hands that milk is given the thousands of forms it can take, expressing the culture and history of a people intertwined with a place.

Skim milk cheeses like pultost and whey cheeses like brunost are an integral part of cheesemaking, representing more holistic, time-tested, resourceful approaches to the craft. They show the inefficiency of treating whey or skim milk as a waste product to be fed to livestock, spread on fields, or put down the drain. All that whey could be pillowy ricotta or sweet brunost. That skim could be pultost or the base for cottage cheese. It takes time and energy to extract these foods, but without them, cheesemaking seems incomplete. That original lesson I learned in Mongolia, that all parts of milk are sacred and should be revered, was echoed in Norway. I heard the melody of that song of the steppe herders, carried by the north wind, emerging from the forests of another land that has not completely forgotten the old ways.

— CHAPTER FIVE —

Refuge of the Old Ways

KRSTENICA, BOHINJ VALLEY, SLOVENIA

In the mountains of truth, you never climb in vain.

—Friedrich Nietzsche

I usually have a contact or a visit planned before I travel somewhere. At other times I simply go with minimal preplanning, taking a leap of faith that someone will appear who can help me. I'll do some reading about a place where interesting traditional cheeses are made, then just go and walk until I find a guesthouse or hiking route that feels promising. An example of this is the time I spent above Bohinj Valley in Slovenia, not far from the Italian and Austrian borders. There is an organized cheese trail that connects various high-mountain grazing areas known as *planinas* that are now offering lodging, meals, and experiences to tourists and trekkers. Like alpages in the Alps or seters in Norway, planinas are mountain pastures where cattle are brought in the summer while hay is grown and harvested in low-lying valleys. It was September, so I didn't know if they would still be open or if I could do much beyond being a paying guest. I rode buses to Slovenia's capital, Ljubljana. After walking around the city, I curled up in my down sleeping bag, stealth camping in a large, wooded park, and took an early train the next morning heading north.

Arriving in the valley's central village, I walked to a tourist information office to see if they had a map of the cheese trail. I told the man working there what I was looking for, and he immediately said, "I know where you should go. I'll call one of the farmers to make sure the cows are still up there." He did so and confirmed that while most of the planinas had closed for fall, the crew at this one would remain for another week or two. He handed me a small folding map and circled a dot. The dot was labeled: Krstenica.

Kristen Nietzsche is how I anglicized the name to remember the pronunciation. I took a bus up to the trailhead, imagining the scowling, owlish face of Friedrich Nietzsche, the philosopher whose words had burned my youthful optimism to the ground at the age of twenty-five. And I'm glad they did. Sometimes forests need to burn, to allow space for new growth. Krstenica. As I hiked, I kept repeating it as a mantra. Saying it now recalls the faith I had at that point in my travels, faith that my feet knew where they were going. *Krst* translates as "baptism" or "cross," while the feminine suffix *-ica* connotes "cute" or "little." Little baptism feels fitting, like a prophecy. For me, Krstenica means that whatever your soul truly needs is out there or in here somewhere. Waiting for you to arrive, to take the toilsome steps out of the valley and into the mystery. Waiting for you to come home to yourselves.

I walked up a mountainside repeating the name in case I needed to ask a passerby for directions. I met no one, besides a few fluttering birds and flighty squirrels. Kristen Nietzsche . . . Kristen Nietzsche . . . I could see it on the map, among a network of trails that allow hikers to visit peaks and have comfortable stays at guesthouses in the painfully picturesque planinas. As I got closer, I saw fresh dung piles along wide paths carved by bovine hooves. I heard cowbells and someone chopping wood. I was slightly apprehensive, not wanting to spoil the tranquility with my invading presence. But there I was. Committed to intrusion. I hesitated, but my heart told my mind to stop worrying: Let our feet do the walking, our hands do the talking. I'll handle the details from the center.

I walked up to a house in a clearing where I saw someone chopping wood and from a distance clumsily said, "Dober don" (good day), one of the few Slovenian phrases I knew. The young man holding the ax looked me over and said hello briskly, speaking in jolting English. His name was Mitja. As we introduced ourselves, I explained to him that I was interested in seeing how they made cheese and that I was hoping to spend time volunteering on a planina. He told me to slow down, to sit, then brought me herbal tea, and began chopping again. I took in the view and enjoyed the nourishing taste of the cool mountain air. Eventually another man came, Marko, who shook my hand warmly, asking me about my background. He had a short conference with Mitja and they told me I could stay and work with them for the final ten days of the summer grazing season.

The planina was a flattish shelf that held a collection of about a dozen small wood cabins and milking barns arranged close to one another like a flock of sheep grazing in wolf country. Some of these were now used as weekend cabins by local families who in the past had come up here with livestock. Krstenica is known for continuing to be an active dairy where cheese is made according to the old techniques, without tourist trappings. Marko showed me to a bunk room in the upstairs of the large main house, and after a quick dinner I fell asleep there, amazed at my fortune. It was surreal. All I did was show up with a willingness to work and a strong desire to learn and participate. Open eyes, heart, and mind, hungry as a spring bear.

The last thing I ate that night was a little cup full of thick sour milk, something now very familiar to me. *Kislo mleko* translates as "sour milk," and it is made by adding a spoonful of a previous batch to fresh milk, which is poured into little cups and left to ferment in a warm place. After twenty-four hours, the milk has coagulated into a slightly sour gel that can be refrigerated until eating. This is a great example of clabber, the kefir-like sour-milk beverage and natural cheese culture in which the bacteria indigenous to raw milk are allowed to ferment and then carried forward in the next batch. This culture born from raw milk becomes a sourdough-like starter that can be kept by a cheesemaker, carried forward through the regular re-culturing known as feeding. Just as cultivated plants adapt to the conditions in which they are grown, natural starter cultures respond to the conditions of the milk, the procedures and ratios of feeding, and how they are stored. At Krstenica, I was baptized in the simple magic of milk microbes, and I would end up naming my own cheesemaking school in honor of my transformational time there. Sour Milk School, at which I am a student masquerading as the teacher. The real teachers are the livestock, the hayfields, the streams that fill the trough, the beings that inhabit the milk.

The Inquisition

Marko is a brewer and anthropologist who looks at brewing the way I look at cheese. He taught me a lot about preindustrial beers, which often involved mythology and ritual, and served as vehicles for medicinal and psychoactive plant compounds. He introduced me to an intriguing concept that became a part of my evolving philosophy. He feels the valley of Bohinj has maintained some of the old ways in terms of not only dairying but also pre-Christian pagan beliefs that continued to exist under the veneer of Christianity. For example, all the houses had a little shrine with a cross and a painting of Jesus that Marko believes was formerly occupied by a local god or saint. These shrines are surrounded by dry plants, many of them locally harvested medicinal herbs. This folk knowledge of healing plants and the worship of local deities and nature spirits is something Christianity has always tried to repress. The imposition of this new religion was backed by imperial powers who, throughout history, have sought to bring the self-sufficient mountain people down to the valleys where they could become productive economic units serving the war machine.

This dynamic continues today. Powerful forces do not want us to grow our food, pick our own medicines, and build resilient communities. He spoke about the **inquisition** as being not just about religious views but about these larger factors. He used the term broadly, as a metaphor for the drive to control life and subdue the human spirit. This drive is apparent in many social and economic forces. According to Marko, the historical Inquisition never ended. It just diversified, hired a PR firm, cleaned up its image, and rebranded.

This insight strengthened my belief that I could find the cheeses I was looking for by visiting places where pastoral people had maintained fragments of these older mythologies, lifeways, and their corresponding food sovereignty. It would be in remote, hard-to-reach lands that had seen less development and were not as desirable to large-scale agriculture or industry that I could find what I was looking for—the borderlands, deserts, and mountains, remote places on the edge of wildlands. The edge is where the action is.

Where the old breeds are kept and the traditional milk foods made, the old tongues spoken and songs sung, there may be sheltered deep wisdom, vestiges of other ways of living with livestock and landscapes that are relevant for those of us hoping to pull out of late capitalism's death spiral with next-day delivery. The cheeses and techniques of milk fermentation hidden in these cloisters can be seen as symbols of cultural resistance that have weathered the waves of empires and the multiple masks worn by the inquisition. Perhaps what I was working against was the inquisition of pasteurization, commercial starters, and "improved breeds," which sees more established biocentric ways of making cheese as being dangerous witchcraft, a threat to be suppressed. We can't, after all, have people growing and fermenting their own food and alcohol, supporting themselves, making their own political decisions without the guidance of tax collectors and priests. That would set a bad example of what real humans living in intact functional cultures look like.

Milking and Making

I woke suddenly the next morning in a dark room, forgetting for a few seconds where I was, with that lonely feeling of being far from the familiar. I heard muffled voices from the kitchen below, smelled woodsmoke and coffee. My worry instantly melted into comfort. Grinning like a chimpanzee sitting on a pile of bananas, I pulled on my cold pants and went down for a quick breakfast of bread, butter, jam, and black coffee. It was still dark and frigid as Marko and I went out under starlight to bring the cows in for milking. The prior evening, they had been milked and released just as it was getting dark, so they wouldn't travel far before bedding down for the night. I had sat with Marko, listening as they headed out onto the mountainside in the last glowing shreds of dusk. As dawn began to stretch its arms across the sky, we took off in the same direction, Marko explaining that it was best to catch the cows before they got up and started wandering off in search of the breakfast buffet. We heard a bell sound lazily from uphill, giving away their location. We headed up and found them perched on little shelves, gazing languidly across a stupendous vista as a burning star promised to start thawing the night air. They rose and stretched slowly, seeming to understand the drill as they sauntered down with minimal prodding, back toward the planina hamlet where a shroud of woodsmoke completed the fairy-tale ambience.

There were two milking sheds built in the same style as the cabins, and the cows entered them single-file, like factory workers clocking in. A vacuum pump was turned on, providing the suction on a simple setup where hoses were attached to teats and the milk pulled into a milk can. Milking proceeded quickly if the cows went to their customary stall, and every step of the ritualized process was performed scrupulously. Dairy animals love regularity, predictability, everything being in its proper place. An unfamiliar jacket slung over a fence is a novelty that may cause a mutiny.

We carried buckets full of milk into the creamery, where Mitja was getting things ready for the day's cheese. There was no bulk tank for cold storage of milk. For thousands of years, milk has been transformed daily into cheese. Refrigeration is a very modern phenomenon that has heavy costs and serious downsides. The nights were cold here and the evening milk was left in the creamery overnight, allowing the cream to rise. This was skimmed off by hand and placed in a larger pail of cream from previous milkings, which was kept in a pantry built off the back of the creamery, partially in the hillside. It felt like 55–60°F (12–15°C) in the stone-walled room where vegetables, meat, and beer were stored. In this cold space, the cream became only slightly sour, which for me leads to the best sour cream and butter. As for many fermentations, learning to restrain the process and not let things go too far is the trick.

Every three days, the cream was made into butter in an old-fashioned wooden plunger churn hooked up to a motor that does the work previously performed by human arms. I only got to taste this butter a few times, as it was rightfully considered the highest treasure of the planina and was sent down to the families who owned the cows. This ethereal butter was profound, likely due to the diet of the cows, the slow, restrained fermentation of the cream in the pantry, and the slow churning in the wood vessel. The distillation of the land into milk and cream was brought into clarity by a whisper of fermentation, the microbiota of the milk interacting with those of the creamery, the buckets, and the pantry—coalescing into a sacrament that allowed direct communication with the spirits of the forest. The fat of the land diffused through my body and spoke to my soul, revealing to me just what butter can be, when we let it. This butter was all the proof I needed that the old gods of the mountains still exist, hidden in plain sight, waiting for us inside a lump of compacted sunshine, laughing at our folly and forgetfulness.

The full-fat, udder-warm morning milk was weighed and poured into a copper vat with the colder, skimmed evening milk. Several starters could be used, depending on what was available. A bit of kislo mleko could be used as a natural starter culture, adding a thriving population of milk-fermenting bacteria to the sweet, unfermented milk. If this was not available, they would use store-bought yogurt, whose bacteria would have originated in a packaged DVI starter culture, the ubiquitous things I was traveling to remote places like this to get away from. They use a powdered FPC rennet, the 100 percent chymosin stuff I was also opposed to. I don't come to these places as an evangelist to judge and tell people they are doing it wrong; I come to watch, listen, and taste. And the milk foods here tasted like the tears of God, mixed with his sweat, some stardust, high-mountain honey, and snowmelt running through a meadow fragrant with recently hatched herbs and birdsong. That milk was goodness incarnate; it conveyed something so old and powerful that the intruding imperial rennet and domesticated yogurt cultures were overwhelmed by the barbarian magic of the hills, the healing plants, grinning rocks, talking rivers, horned cattle, and dancing pagans who drink ceremonial home-brewed beer to honor it all. The inquisition may be strong, but the people and spirits of the hill are stronger, older, and in it for the long haul.

The curd was cut with a harp like the one used at Arpisson, then cooked to the exact same temperature as tipo fontina: 113°F (45°C). Many aspects of the process are similar, because these cheeses follow the thermophilic technique that was introduced to both areas. This technique requires a kit of tools: the copper vat, a wire harp for cutting, adjustable bands for shaping large wheels, with pressure applied to get them to fully seal up, preventing intrusions of mold into the paste. The technique is based on cutting a delicate rennet curd very small, then cooking it rapidly to a high temperature. Once you get above 104°F (40°C), you begin encouraging a category of **thermophilic** (heat-loving) **bacteria** to grow, which impart particular flavors to a cheese as it ripens. These are often sweet flavors of caramel, brown sugar, or maple syrup, along with nuttiness and savory, meaty, brothy characteristics. Popular thermophilic or "alpine" cheeses include Gruyère and Comté. The very hard Grana styles of Italy such as Parmigiano also represent this approach and display thermophilic flavors, as they are cooked to temperatures above 120°F (49°C).

The technique results in a narrowing of the microbial community in raw milk and selects for those bacteria that are tolerant of this heat, including the thermophiles that are used to make yogurt. Most microbes are inhibited at high temperatures, so the thermophilic technique minimizes the growth of many microbes, including potentially problematic ones. By quickly cooking sweet milk above the **mesophilic** (middle-loving) temperature range, you are steering toward a fermentation dominated by thermophiles. This leads to a more consistent and predictable cheesemaking process, which is especially important with these large wheels that are meant to be stored long-term, transported long distances, and used as a trade good.

There are also less firm, washed-rind alpines such as raclette and fontina that are not cooked to such a high temperature. They are cooked up to temperatures in the range of 104–115°F (40–46°C), following the thermophilic technique but not going all the way. I feel that they are not exclusively thermophilic, and the flavors are somewhere in between meso and thermo, containing aspects of both. Further down the spectrum are cheeses like cheddar, which are often cooked to 104°F, just inside the thermophilic realm. In such cases, the fermentation is mainly mesophilic but thermo starters are often added as well, and the heating wakes the bacteria up enough that they multiply and will add their flavors as they die and release enzymes inside the ripening cheese. Like most dichotomies, the one that says cheeses are either meso or thermo is often false. There are a range of hybrids out there. There is an in-between zone, a blurry border region, where some very fascinating cheeses dwell, with feet in both worlds, transcending the boundaries that humans attempt to impose on the natural world.

The meso-thermo alpine cheese that I describe is sometimes referred to as Bohinjski sir (Bohinj cheese), but I think we can get even more specific and call it Krstenica sir. The cheese made at Krstenica. Once it had barely firmed up, the curd was cut rapidly and a fire placed under the copper vat by a clever movable fire pan hidden below the floor. It was cooked while being stirred constantly, then removed in a couple of large masses. These were carried to a wooden pressing table and broken up to fill molds; wood lids were then stacked on top of the warm cheeses, pressing down on them. Pressure was applied from an overhead cantilevered wood beam, the same setup I had seen in a museum dedicated to the dairy traditions of the valley. The cheeses were pressed and cooled overnight before being salted in a brine tank and then aged for three to six weeks in a semi-cool, closet-sized aging space. They develop flavor rapidly, with notes like tipo fontina: olives, cured meat, and sourdough bread. This cheese doesn't go through the salt washing and smear development of fontina but has a very diverse flavor profile of yeasts and the maltiness of cheeses made with clabber and kefir starters.

The Deep Terroir of Krstenica

There is a minor cast of milk foods that accompany Krstenica sir and planina butter. Most of these have essentially disappeared, but Marko was making them for his own amusement, out of an appreciation for their role as food and medicine for hardworking herders. One that is still common is a ricotta called *sladka skuta*, made by heating whey to nearly

200°F (93°C) and adding a hefty dose of citric acid, causing a dramatic separation of faintly gray ricotta and clear, green whey. Compared with the approaches I've seen in Italy, more acid and heat were used, leading to a higher yield of drier, less delicate, less delicious cheese. Marko was making an aged rendition of this called *slana skuta*, in which the ricotta was salted heavily to what tasted like 2.5 to 3 percent and packed into a bowl or barrel with cheesecloth on top, along with a wood plate that was weighed down with a heavy rock, pressing out some whey. The cheese was kept submerged in this salty whey, as if it were sauerkraut. The vessel full of salty ricotta and whey would be left close to the fireplace, where a range of ferments were kept warm. After two weeks, there was a red smear on the surface of the whey that my nose confirmed was a community of salt-tolerant bacteria, similar to what develops on smear-ripened cheeses. Meat, broth, cow barn, and dirty socks. Thankfully in that order. The top layer of the cheese had a reddish hue and suggestions of the smear in its flavor. Other sections further down the sides tasted like blue cheese. The center was much less funky, tasting like skim milk feta. It was an impressive range of flavors in a two-week-old brined ricotta with a texture very unlike the fresh cheese.

The scotta remaining after the ricotta is ladled out is called *kisava*. Now it goes down a drain and out to a trough for the cows. In the past, it would likely have become food for the hogs that were brought to the planinas to be fattened. A small amount could be saved and left in a vessel near the fireplace, where it would sour to be used as a health tonic, or as the acid to make ricotta. Marko had a plastic bucket of it that was months old, and it was nearly as sour as vinegar. It also had a floating biofilm called a **pellicle**, like the cellulose cap seen growing on top of kombucha. This had orange-red smear bacteria growing on it and smelled like the aged ricotta, indicating a common SCOBY in the environment. Where was this microbial community coming from? The plants and pastures? The bodies of the cows themselves, and the grassy knolls where they slept? Was it living on the walls and floors of the creamery and where milk was processed daily? Could it be in the air or water, or on the bodies of the makers, somehow given shape and coherence by the steering of a microbial potential via their tools, recipes, and techniques? Maybe it was little bits of all these things, and maybe we will never be able to untangle the complex web without breaking it in the act of dissection.

Krstenica sir and most of the fermented dairy foods made alongside it displayed a common background personality. They exhibited a shared terroir, a unifying aromatic signature that I also detected emanating from spaces on the farm. I first noticed it in both the cheese and the thick sour-milk beverage (kislo mleko) served at breakfast and dessert and sometimes used as a cheese starter. Then I noticed it coming from the body of a cow and permeating the wooden posts of the milking barn. It was a yeasty sour-milk and cow aroma, with aspects of the aromas I associate with naturally fermented cheeses. It was these, but combined with other traits I can't contain in words, the sum of which was one of the most distinct instances of deep terroir in cheese that I have experienced. I clearly detected this aromatic signature when entering the creamery. I found it emanating from a wooden table where the cheese was pressed. The cheeses, still in the early stages of fermentation, sat overnight on this wood press as they cooled, being "contaminated" by a biofilm that was thriving on a porous surface that could never be sterilized. The web of connections expanded, with the wooden press serving as a nursery for the yeasts that grew readily on this cheese, the equipment and room itself being a vessel for a culture that was being carried forward daily, by keeping this culture fed with milk, whey, and cheese.

The milk itself had a mother personality that was expressed in her lineage of children, who embodied aspects of their mother in various ratios, having transmitted her spirit via different pathways of growth. This personality is born on the hillsides where the cows graze and sleep, from the spring water that fills their wooden troughs, and from the unseeded pastures they graze. It is transformed in their bodies into milk, which is received by humans in an old wood structure with dirt floors and cobwebs that could never be sanitized. The milk travels from this structure about fifty feet to the cheese vat, where the same hands that just milked the cows pour the warm liquid into the vat to begin the next phase of its transformation. No attempt is made to isolate these functions; there is a continuity between the fields, the milking parlor, and the cheese vat. Cheeses can be unique expressions of the place they are made when we allow this continuity to occur, when our cultural practices amplify rather than silence the personality, the voice, the wisdom of the land. When we stop shouting orders and remember how to listen.

— CHAPTER SIX —

A Cheese Called Queso

JERTE VALLEY, EXTREMADURA, SPAIN

Men rush towards complexity; but they yearn towards simplicity. They try to be kings; but they dream of being shepherds.

—Gilbert K. Chesterton

I rode buses from Madrid to the north of Extremadura, a dry inland region of Spain bordering Portugal. Jerte Valley is high and relatively well watered, so fruit trees are grown as a primary crop. I had read an article in *Atlas Obscura* titled "The Last Goat Herd in Hell," about a man named Alfonso Hernández Salguero, who was continuing to raise an old goat breed and graze them on the rugged terrain above the valley, which is famous for its cherries. I emailed him in Spanish, and he said I could come out to stay for a few days.

I got off a bus on the side of a quiet country thoroughfare and started walking up dirt roads toward a hilltop town high above called Piornal, where Alfonso had told me I could find him. The hills were cut into stone terraces covered in cherry trees. I started finding all sorts of other fruit trees and vines along the roadside or on the edges of terraces. I gleaned a bunch of purple grapes, figs, and apples as I climbed, feasting on the abundant unfenced food forest with no humans around to tell me not to. As the light retired from the sky on this hot October day, I wandered off the road, passing decaying stone barns full of cobwebs and rusting farming equipment. I found a nice perch where I felt safe and hidden from sight, and set my sleeping pad onto my trusty poncho-tarp tent that also served as a ground cloth. Dinner was three top-notch Spanish sheep cheeses, a stick of chorizo, and all the fruit I had picked on my short walk in a new land.

Waking to a light dew, I made my instant coffee in cold water and started walking uphill to warm up. The list of perennial crops I found interspersed in the cherry orchards grew: persimmons, apricots, pears, quinces, almonds. About halfway up the tall hill, it got steeper and cherries gave way to olive trees and trellised vineyards. I went through an eerily quiet village that was semi-abandoned, like many in this depopulating agrarian region. Signed trails led out and up, and as I climbed, I found little vegetable gardens of peppers and eggplants sprinkled among fig and apple trees, with a riparian corridor left alone for thickets of brush to grow below alders and willows, a sanctuary for birds and all sorts of critters. The orchard crops and terraces ended at a certain elevation, giving way to oak forests and then groves of ancient chestnuts. This is another important crop here, both for the nuts and for the dark, musky honey that is prized in Spain and Italy. What appeared to be perennial water flowed down stone-lined channels and was spread out onto the lush terraces below. This is one of the more impressive permaculture arrangements I have seen, but I doubt the last people left living here called it that. It was simply the result of generations of people living on the land, figuring out what works and what doesn't. Following common sense over an extended time frame. Moving soil and shaping the earth in a smart and respectful way, with small-scale terracing and water-harvesting swales. The land was pumping out food from trees and vines planted long ago that perhaps hadn't seen much care in recent decades. A permaculture ghost town, longing for the return of laughing primates who can repair stone walls and facilitate continued abundance for the whole living web.

Above the oak-and-chestnut layer, the plants shifted to arid Mediterranean chaparral that looked similar to parts of Southern California, like the hills above San Diego. Sages, ephedra, woody herbs, and low-growing thorny scrub sat between huge boulders on gravelly soil, with patches of flat grassland appearing as the hill leveled out on a rocky ridge. I was in touch with Alfonso via WhatsApp as I entered the town of Piornal. Shopkeepers stared at me and scurried inside as I sauntered down the center of the street like a cowboy looking for an open saloon. He sent me his live location; a tech-savvy goatherd I was dealing with here. His location kept moving, but I eventually caught up, turning off a dirt road as I heard a soothing symphony of bells clanging from off in the brush. The original, auditory GPS: goat positioning system.

Alfonso was a stout man who carried an ax rather than the customary shepherd's staff. He wore a wide-brimmed white hat cocked at a rakish angle and had the air of a flamboyantly masculine Spanish man. He started speaking to me in rapid Spanish, and I had to break the news that I didn't actually share his language. He kind of scoffed and raised his eyebrows in a tired dissatisfaction that seemed to be his default vibe. He turned his attention back to the goats, and walked off in the direction they were heading. I hurried to keep up, wearing a childlike grin as my backpack got caught up in the thick brush, satisfied that I had made it here to walk the land with a proud and forlorn herder. The last of his kind.

The goats were black, with small patches of brown and gray. Some were mainly brown, with a black spine and legs. Others were a speckled bluish gray, with coats that reminded me of blue heelers. They had long horns that slowly spiraled up and away from their heads. The breed is called Verata, but my travels were beginning to show me that in some places breeds are not as fixed and distinct as many conceive them to be. The multitude of coat patterns indicates genetic diversity and a less rigid sense of husbandry, rather than breeding toward a codified "breed standard."

It is a trait of the modern, rationalist, taxonomic mind to want to fix names on things, to put life-forms into neat boxes where they will stay put. All livestock have to fit into named breeds, microbes have to be fixed with a genus and species, cheeses need to belong to a style. This mindset can't handle fuzzy boundaries, spectrums, or shifting, transitory states. It despises unchained diversity and things that can't be pinned down, identified, and cataloged. The way the pastoralists I met referred to their animals was often not in terms of definite breeds, but in terms of a place. "These are the goats we've always kept here," Alfonso told me, rolling his eyes at my continual attack of questions and walking quickly after the goats. The goats of Piornal, which his family has always kept.

He was continually vocalizing to them with loud whistles, shouts, and unique calls. An ascending whistle seemed to mean "keep eating, you are fine, I am here." Louder, more abrasive shouts were directed at wanderers or lingerers, telling them they'd better get back into the group. These sounded similar to dog barks, and I marveled at the possibility that human herders may in fact be imitating dogs and filling a similar role in the mind of the goats. A third call was used to bring the goats toward the herder, and was associated with getting treats like a pellet grain supplement or a chopped oak branch covered in leaves. It was a descending "DAHHK, DAK DAk dak dak." There was also a loud, popping two-tone sound made out the corner of the mouth, kind of like when someone on horseback gives two loud clicks. The whole family was capable of these same vocalizations. I tried to mimic these sounds, and the goats just stared at me incredulously, as if I had asked them a highfalutin philosophical question. It was the same look Alfonso gave me when I asked him just about anything. I figured I would just walk with him and the goats, watching and listening. From then on, the lessons poured in.

There were about three hundred goats, including intact bucks that would cruise around searching for does who were in heat, sticking their Elvis lips up in the air. There were many pregnant does, and each day at least three of them gave birth. When this happened, Alfonso would wait to make sure the mom was cleaning the fresh kid, then we would continue with the herd, which never seemed to stop moving for more than a few minutes. I am used to seeing goats and sheep give birth in a barn and being hovered over or assisted until the babies are standing and drinking milk. Like so many things I would see in places with a deep history of dairying and keeping old breeds, the birthing process in Piornal was much more hands-off. In the evening, after the herd was back in their night pen at the home farm, we would drive into the fading dusk, back to the places where we had left the new mamas. Sometimes they had moved a bit and dropped another baby. When we couldn't find them, we sat and listened, eventually hearing a mama rustling in the brush or a baby bleat. I delighted in catching the little ones, who were already running around and able to evade me and other predators. We would carry the kids by their front legs, bodies dangling down. They screamed out, and the mothers followed anxiously, calling back as worried mothers of any species will, in an unmistakable language. We put the babies in a small trailer-turned-nursery and loaded in the moms, then went to find the others.

Among the common traits of heritage breeds are ease of birthing and good mothering ability. The mothers have fewer complications requiring intervention when giving birth, and very rarely reject or lose their babies. They are good moms who know what to do; they still possess their inherent animal psychology and biology. Somehow, these inherent capabilities can weaken when modern breeding programs select intensely and exclusively for traits such as high milk volume or faster growth and larger carcass size. As so many herders and farmers who keep the old breeds told me, when you breed for one trait, you lose some of the others. Modern breeding has led to unforeseen frailties, to overly domesticated animals reliant on imported feed, medicines, and supervision. Modernity's desperate rush to take everything spontaneous, organic, and nonhuman and force it to fit the needs of sociopathic economic systems has resulted in many profound losses. Our animals, foods, homes, landscapes, and bodies have lost the vitality of living things, the will to live being replaced by a facade of efficiency in which this devolution is referred to as progress.

In the eyes of these goats, I saw another way of being, something between wild and domesticated. I saw the strength and spark of what is indigenous and alive, in animals that had joined up with humans long ago, that had been changed by us into something not wild, but also not made subservient and degenerate. They were tame but maintained the dignity and prolific vitality of their true goat nature. These goats and this landscape are a testament to how humans can breed animals and impact landscapes in a positive way that benefits all life. This is what we do. We're just forgetting how. We are under the influence of a powerful spell that does not want us gazing into the eyes of a goat and recalling our deeper ties to the beating hearts of our lost herd.

We need these goats as much as they need us.

The home farm was surrounded by a sacrifice zone, bare dusty soil and rock that had been hammered by this large herd for who knows how many decades. The typical detritus of a dairy farm lay around haphazardly: tarps covering aluminum racks full of plastic tubs, boxes of dirty spray bottles, fencing equipment, skeletons of half-finished projects, random defunct appliances. A single-room concrete building was the home of Alfonso's mother and father. The scene was one of poverty, a common reality that can easily be left out when celebrating the more picturesque and admirable aspects of the lives of pastoral people. A smoky fire burned in the corner, plastic bags full of stuff lined the walls, the corners were piled with the same crap in boxes that agrarian hoarders always have. I sat at a small kitchen table while Alfonso and his parents yelled at one another. They appeared to be fighting and it was incredibly awkward to just be sitting there, not understanding the words. It can be a fine line I ride, between being fortunate to be welcomed into someone's home and being trapped in a remote place with strangers. Suddenly the vibe shifted and they smiled and laughed, the whole scene was seemingly routine to them. These were isolated folks, living up in the hills with little social contact. And they probably liked it that way. Despite their rough nature and the air of frustrated exhaustion, they hosted me with the default hospitality common to all of the pastoral people I have visited. A bed was made for me on the floor, with a thin mattress upon which I laid my sleeping bag.

Alfonso's dad was a short, wiry man in his eighties whose back arched with extreme kyphosis, certainly due to a life of hard labor and milking. I've seen many old lifetime dairymen with the same question-mark shape, still out chasing cows and fixing tractors with the same determined momentum ranchers often move with. His mom was a whippersnapper, and she laid into Señor with a gruff vitriol that slowly bubbled up on her face, then erupted in a voice hoarse from decades of yelling and fighting. She cooked all our meals and did all the cleaning. The men did none of these things, unless there was meat to be grilled. The cuisine was real, the no-nonsense, hearty home-cooked meals of rural peasant farmers. That may sound romantic and stereotypical, and it was. It was also resourceful, and full of local seasonal produce. One night we ate fresh broad beans boiled in their pods with potatoes and red pepper, wild hawk's wing mushrooms, and chestnuts roasted on the fire. In the middle of the table was an infinite pile of bread.

There is a focus on bread in many peasant and pastoralist cuisines. It's cheap, transportable, and versatile. You make a bunch of it, then have food for days. They would get huge bags of beginning-to-stale bread free from a nearby bakery. Some of the most comforting dishes or ingredients in the world are made from old bread. Bread salads, puddings, and dumplings. Strata, *ribollita*, stuffing, French toast, croutons, breadcrumbs. Here they made a soup called *sopas canas*. It started the way many things do in Spain: with a lake of olive oil in a hot skillet, garlic, and *pimentón*, a smoky red chili powder. Once this was highly fragrant and the garlic was golden, goat milk was poured in and reduced for fifteen minutes. Then chunks of bread were tossed in and cooked until they disintegrated. More bread was thrown in and the pot taken off the heat. Served in large bowls, the soup contained multiple textures: wispy, starchy strands from the dissolved bread and varied soggy to crunchy hunks from the later addition. All in a thick, smoky paprika-and-garlic milk broth. That soup was soul feeding, a love poem in a bowl. I will never forget it, and have made it many times since, never quite recapturing what I had in Jerte. These simple recipes with few ingredients are often the hardest to replicate. And when you're trying to match grandma magic, you might as well give up.

Comida (lunch) was the biggest meal, with two courses followed by dessert and coffee. Once, we had colostrum, sweetened with sugar and thickened to a pudding on the stovetop. Colostrum is kept separate from milk, as it defies coagulation and can be problematic in cheesemaking. Many dairying peoples will milk out colostrum after the baby has gotten its fill and use it for a baked dish similar to this, or freeze it for use as a cure-all. After a lighter *cena* (late dinner), we could help ourselves to a large container filled with cheese and cured meats such as dry chorizo, soft *morcilla* (blood sausage with rice), and *jamon* (ham). We would also take snacks with us while walking with the goats. Bread with lots of meat and cheese, one little bottle of olive oil, and another of wretched gut-rot homemade wine.

It was kidding season when I arrived in early October, before the winter rains had started to fall. The moms and their kids spent the night together in a barn that looked like a haunted house. A dance of decay, repair, and disrepair was painted on the walls by the slow, steady hand of time. Time may heal all wounds, but it also erodes all walls, and this barn perfectly embraced this reality. We woke well before dawn to catch the does still bedded down with their udders full, before the kids started nursing like psychotic milk vampires. The moms were milked wherever they stood; it felt very chaotic. By the end, the whole family was chasing the few that still had full udders, cursing at them in a routine they repeated daily. There were over one hundred kids and they caught each one and made sure it drank until full, from either its mother or a surrogate. They were then placed in a daycare zone with a condition known as milk drunk in which the babies stagger, hiccup, and shiver, with a spaced-out look in their eyes, then pass out. The milk was placed in a bulk tank powered by a generator. It was sold to a co-op creamery to be made into cheese, likely by mixing it with the milk of many farms and pasteurizing. They only kept a small amount for cooking and making their own cheese.

A bit after noon, Alfonso and I would start walking with the herd down miles of dirt roads, through open oak savanna studded with patches of thorny scrub and rocky, hilly sections covered in prickly chaparral. The goats fed on a range of plant communities as we did a grazing circuit, circling away from the farm and around other properties, visiting a spring where water flowed between ancient stone troughs. This was a landscape manicured by the goats, and while their impact was obvious, it did not appear overgrazed. In wildfire-mitigation planned-grazing efforts, there is a practice called breaking the fuel ladder, which I recognized in the less rigidly planned grazing patterns of this herd. The majority of vegetation on the trunks of trees was pruned from the ground to about five feet up, as high as goats can reach as they browse vertically. This gap would allow wildfires to burn quickly through a ground level of short, dry grass and dead branches, while preventing them from climbing into the canopy above. Goats, sheep, and other herbivores are a crucial component of maintaining fire-resilient landscapes in our increasingly fire-prone future. With fewer people and animals on the land and increasing aridity, Spain is on the front line of this future that is already here.

On one of my final nights, I went with Alfonso to his home down in the valley and was surprised to learn that he lived in a nice modern house with his wife and children. We took gallons of milk with us and heated it in a giant pot over a propane paella burner, a common setup in Spain. There was no added starter; we just poured in rennet and let the milk coagulate. Alfonso used his hand with fingers spread wide to cut the curd, then made a brisk circular motion to cause an upwelling and get the curds moving. He then put his hands side by side and slowly applied pressure to the curd, pushing it against the pot and causing it to knit together in a massive blob. He removed hunks of cheese with a strainer and tossed them into PVC rings on a well-used wooden draining table set at a slight angle so the whey ran off into a waiting milk can. He used his fingers to perforate the wheels in a pecking motion, releasing more whey. The cheeses were salted on one side with a sprinkling of coarse salt. In the morning, they were flipped and the other side salted.

It is uncommon to salt a cheese like this immediately after it is made, but I have seen it done in hot climates where milk ferments rapidly on its own. Salting slows but doesn't fully inhibit fermentation. This cheese was very salty; I would say just on the side of too much. Heavy salting is also common in hot places as a means of preserving cheese without refrigeration or a cool cellar. Alfonso's cheeses aged on the wooden draining table, which smelled of sour whey and yeast. After one week of daily flipping, it got yeasty, no doubt helped by the wood that was soaked in salty whey and held on to moisture and a yeast biofilm, creating a microbiota that interacted with the cheeses. By two weeks, it was beginning to dry and develop a yellowish rind with spotty growths of *Geo*. It was ready to eat and had a surprising amount of flavor for such a young cheese. The paste was firm, bordering on crumbly, likely due to the high acid content of this bright, clean, goaty wheel. Cheeses like this are sometimes coated in olive oil and paprika and aged a bit longer, up to two months.

"What is the name of this cheese?" I asked Alfonso. He looked at me quizzically, as if I were a donkey trying to do algebra. "We just call it *queso*," he said. "It's the cheese we've always made. My grandparents always had it on their table, like Americans have a tub of butter." This is simply, effortlessly—without definition, trademark, or governance—the cheese of Jerte Valley and the ridge above it. The cheese, the goats, the family are all defiantly taking a stand for the way they coexist interdependently. They proudly stand for a lifeway that is rapidly eroding, that they see as right, time tested, and earned through hard work in a world obsessed with the new, comfortable, and mediocre.

A cheese named cheese, that tastes like perseverance and common sense.

I'll never forget it.

What impressed me about this visit was the sense that this family was a vestige of a human-livestock-landscape synthesis that had visibly shaped this place. This watershed-level food ladder was exemplary of the sustainable, sovereign, and nourishing models we seek to build today. It's being abandoned, but the structural wisdom remains, producing copious food—not only for humans, but also for the land and the macro- and microbiology that inhabits it and holds it all together. You can walk through it, as I did, from bottom to top. The goats and their keepers inhabit the top rung of the ladder. This is the driest, most exposed and rocky layer, least suitable to raising crops or planting orchards. In the past, when there were more people living here, the goat herds would be brought down in the fall and winter to clean up the groves of olive and fruit trees, where they would knock back vegetation that had grown tall over the summer as their dung and urine were incorporated as fertilizer with help from their hooves. To get to the lower orchards, they passed through the chestnut and oak rungs, sharing the nuts and acorns with hogs who got fat on these prolific trees. In my journal, I charted this cascading cycle of integrated perennial crops, agroforestry, and pastoral marginal lands—with the goats carrying the nutrients from the bottom of the ladder back up to the top, where the nutrients could flow down, along with the water from the springs, which conveniently spilled out on the ridgetop, providing water for humans and goats before flowing down through the terraced ladder.

The cheese was simply a food resulting from this foodshed, and humbly stood as one plate on the table of Jerte Valley, which in the days past would have included thin slices of acorn-fattened jamon and seasoned chorizo; olive oil, cherry preserves, quince and peach jams, dried figs and pears, persimmons, eggplants, peppers, beans, chestnuts, foraged mushrooms and greens, trout from ponds, honeys, apple cider, homemade wine, roasted goat, chicken eggs, turkey, and wild game. The raising of the goats was not a discrete operation but an integrated, highly mobile component of a dynamic, human-sculpted, perpetual biological engine. The goat herd—a multispecies cohort that included humans, dogs, and horses—could move where needed, its impact enriching rather than stripping the landscape, the manure and urine not being seen as waste but treated as the gold that they are, as food for the soil and its microbes, upon which the whole cycle of life is built.

Alfonso is a stubborn defender of these old ways and this old breed, but he is also fatalistic and emits the aura of doom so commonly seen in surviving knowledge holders of once-vibrant pastoral cultures. I hear it over and over again: "We are the last ones." The feeling is that this accumulated wisdom is being given up, simply dropped and forgotten, because there is no one left who wants to learn it. I asked him to describe how he plans to work toward keeping the old ways alive. He thought it over and replied with conviction, "My job is to care for, love, and keep safe and healthy the animals that care for, love, and feed us in return. The future is dark, but as long as I am doing this, the heritage of my ancestors stays alive." He doesn't want his daughter to take on his lifestyle, feeling she should have opportunities to pursue education and a career of her choosing. Alfonso and countless other shepherds I have met feel there is no future in this way of life. I feel otherwise.

His daily motions and demeanor struck me as somber, a funeral for what was right in the world, what made sense and worked. Yet the seeds that Alfonso holds in his worn hands, that he keeps watered with tears in his tired heart, are priceless. They are the legacy of a people, a way to live in relationship with the holy ground, to feed the place that feeds you. Perhaps without knowing it, Alfonso dropped a few of these seeds into my hand. I added them to the sacred pouch that hangs in my chest, between my heart and head, put on my backpack, and walked back down the hill.

— CHAPTER SEVEN —

The Shepherd Cheese Trinity

SICILY, ITALY

The people are the recipe.

—MAX JONES,
transhumance guide, writer, photographer

The sheep were tinged yellow by dust mixing with the lanolin wax of their wool. They moved like a slow-motion flock of birds across a grassy hill in a hopscotching formation. Some walked slowly in a phalanx with their heads close to the ground as they nibbled and fell behind, then ran to catch up and become the new front line. We kept our distance, and took the pressure off completely as the flock entered an olive grove and dispersed a bit to graze peacefully. When it was time to go, we had to push them forward with hoots, waving of arms and sticks, and shouts of "HOM-INY!" (actually *Amunì*, Sicilian for "let's go"). They formed single-file lines, funneling into interwoven trails that cut through a brushy section crossing a ravine, where they bunched up in a traffic jam at the bridge, which was kind of scary to cross.

This was one of the flocks cared for by the Cangemi family and we were moving them between properties, a *piccola transumanza* (little transhumance). It was March in the coastal hills of Sicily, and the green pastures of a wet Mediterranean winter would soon desiccate and turn gold in the relentless summer sun when plants wither and go dormant in order to stay alive. The milking season typically starts in December and winds down in midsummer, an inversion of the norm farther north. The sheep were being moved to a lower property with its own milking shed to allow the pastures around the home farm to rebound and serve as a store of dry feed later in summer, when the whole herd would be brought back up.

The sheep were a heritage breed known as Valle del Belice, named after the valley in the province of Trapani where the family has farmed land since the 1700s. They looked quite similar to the Langhe sheep, with coarse, white wool and strong, bald legs built to handle long walks on rough terrain. In the past, sheep would lamb between December and February, in sync with the greening up of pasture, then would be dried off in July and August after the wheat harvest, when the pastures and croplands try to take a siesta and there is little vegetation around to provide the nutritional power to make milk. Sheep's milk is increasingly produced year-round by staggered breeding. In an attempt to stay afloat in a dismal economic situation, many producers have taken on larger flocks and are feeding a high-calorie grain diet, forcing animals to produce as much milk as possible so cheese can be made regardless of the season or what is best for animal health and the land.

The sheep packed together, heads down, as a mob that funneled into lines when they neared the small ramp that took them up to a seated shepherd. A head gate held the sheep in place while they were milked out quickly into a plastic bucket. As the rookie volunteer shepherd nicknamed "Americano," I was tasked with standing behind the flock with a staff and pushing them into the milking lines to apply steady pressure on those in front to walk up the ramp. I spent many hours standing there, watching the sheep and observing how each shepherd interacted with the flock. Giorgio calmly cooed and purred to them in soothing bass tones, never losing his cool or making sudden movements. He had obviously spent a lot of time around sheep, and his approach was effective. Millete was less calm and struggled with the flock. He grabbed them suddenly, and yelled at me when they were apprehensive about approaching his more agitated energetic field. Davide Cangemi, born into this shepherding life, simply waited when a sheep hesitated to enter the stanchion. His cell phone sat close by, and he texted or swiped with an air of resigned boredom. Looking straight at a herd animal (especially sheep) and singling them out can cause stress. When animals are stressed, it's hard to work with them. It's often better to look away and almost ignore them like Davide did, letting the mood relax.

This was the first leg of my travels with a fellow cheese explorer, Carina Reckers from Germany. Joining forces, we explored some of the cheeses made on the island and the culinary traditions in which they nested like meatballs on spaghetti. A cheese eaten outside of the cuisine from which it sprouts and dropped on a plate in a foreign land lacks something, like picking up sushi and a side of trail mix at a gas station in the middle of a desert, and eating it while driving on the interstate. The uprooted cheese is out of context. Nebulous. Homesick. Missing its friends, the drinks and dishes with which it comfortably shares table space. To get to know a cheese, you must see the place it is from, visit its home, meet its parents.

We spent two weeks with la Famiglia Cangemi, learning by participating in their work as volunteers. Carina worked with the women, in the caseificio. I worked with the men and sheep, in the fields and milk sheds. This division of labor along gender lines was part of the conservative Catholic societal norms of this backwoods part of the island. We stayed with Davide and his fiancée, Martina, in the living room of their small cinder-block home that looked like a trashed farmhouse anywhere in the world. Same broken shit, continual maintenance, fencing and vehicles in a continual flux of neglect and repair. Beer bottles, cigarette butts, and too many puppies. A crucified Jesus watched over us as we slept uneasily on a foldout bed. Every morning, Davide and I had coffee before dawn, then rounded up Giorgio and Millete, who were hired farmhands. In the frozen, hazy light, we packed into a truck and drove through citrus groves, down to the lower milking area where we had walked the sheep. On the way, we ate cheap cookies or other sugary junk food. "No sugar, no milking," Davide told me. He had me roll up a long cigarette seasoned with hash as we drove, the first test of my worth as an aspiring shepherd. As the sun rose, while the shepherds brought the milking herd in from their night pen, Davide would search his pockets for the lighter—he did this every day—then ask me for the *accendino* with the universal hand sign.

"No fire, no party."

As we started milking, Davide's older brother Calogero would arrive. A photo of their grandfather, Nono Calogero, pinned to the wall, watched over the process. As the eldest son, Cali was the family's front man. He always had clean new clothes and a pristine, fashionable hairdo, and he wore cologne. Carina called him the *pastore profumato*, the perfumed shepherd.

Like most people tasked with milking a few hundred sheep by hand, they went fast, handling each sheep for perhaps ninety seconds on average. Some barely had any milk to drop. There was no teat washing, stripping, or iodine dipping. The sheep were kind of scruffy and dreadlocked; many had dirt and poop dust in their wool. A lot of this debris would fall in the five-gallon bucket held between the shepherd's legs, directly below a sheep's butthole and woolly dingleberries. Once the bucket was nearly half full, they would pour it through a mesh filter into tall plastic milk cans. The filter caught a layer cake with a thin panna cotta of poop-dust sludge as the base. Above this, a thick wool cookie was topped with flecks of hay and plant matter garnished with sheep turdlets. I was mildly sketched out, but also, paradoxically, hungry for dessert.

When we were finished, Cali or one of the shepherds would take the sheep out for their daily grazing circuit. The flock grazed on a mixture of scrubland, rocky hillsides, fallow ag fields, and olive and citrus orchards. The area was a true mosaic of land uses, with wheat being the major crop grown in flat areas, then vineyards, olives, or rangeland filling in the more rugged terrain in varying bands as hills rose into uncultivated mountains. After wheat was harvested, the straw was reserved as winter fodder for the sheep, who also came into the wheat fields post-harvest to feed on stubble and fallen grain while depositing manure and urine as fertilizer, perforating the fields with their hooves. Textbook agropastoralism. Straw is the low-protein stalks remaining after the grain is removed; feeding it required supplementing with richer hay cut from river bottoms. These loosely cultivated hayfields had a diverse blend of flowers, herbs, and an alfalfa-like legume known as *sulla*. Sulla grows wild on the island, and is also cultivated in these hayfields, as a low-input cover crop that can be grazed or cut. The long history of sheep dairying on Sicily is deeply integrated with the growing of grain and tree crops such as carob, citrus, and olives. The movements of the sheep are still tied to the growing season of wheat, which has been a major crop since 7000 B.C.E. How could this soil remain fertile after producing grain crops for this long? Animal poop. Without livestock, sustainable annual crop agriculture is a much harder feat to accomplish.

The milk was now sitting warm in plastic jugs, basking in the uncompromising Sicilian sun outside the milking parlor. We loaded up the truck and drove the milk to the caseificio. A lot of the sanitation protocols that I had been taught were necessary in a creamery did not exist here. We walked into the caseificio with our farm boots on. A sales counter and cold case were in the same space, and the customers walked right in, with no defined separation between the public and the production area. Much of the equipment was only washed with hot water. Not only was the cheese safe to eat, but it was also delicious in a way that can't be replicated with commercial starters, HAACP food safety plans, modern "clean" milk, and impeccably sanitized stainless steel. When cheese is that clean and controlled, it usually lacks soul. The cake of shit caught in the filter made me think the cheese would taste bad. But the cheeses were often spectacular, and made without adding a starter. I remembered what I had learned at Cascina del Finocchio Verde: The starter culture is already in the milk.

The milk was born from healthy livestock eating plants growing from living local landscapes, producing milk thriving with healthy microbes. The proliferation of pathogenic microbes in cheese is related to imbalances in milk created by broken food systems. Healthy soils are not where we see plants with fungus damage, insect infestations, or the spread of invasive species. These are symptoms of unhealthy farming practices and soil ecology degraded by monocultures and mistreatment. It is damaged, degraded ecosystems that are susceptible to pests or fungal pathogens. Invasive species are a symptom of the misuse of landscapes. Dangerous, invasive microbes are a symptom of damaged, degraded milk coming from mistreated animals. The dangers of raw milk are created when we turn it into a commodity, when we are guided by a pathological vision that sees life as something we should transform into money rather than worship.

Pecorino Siciliano

The creamery produces a trinity of cheeses with different shelf lives. Pecorino Siciliano is a hard DOP cheese, one of the stalwart workhorses of Sicilian shepherd cheesemaking. From the same base curd as the pecorino, a mozzarella-like cheese is made that is eaten fresh, up to two weeks old. Called Vastedda del Belice, this is a rare sheep's milk stretched-curd cheese, molded in a soup bowl. The whey remaining is enriched with reserved milk and made into ricotta. This is the super fresh cheese made and sold every day (besides Sunday) to bring in immediate revenue. These three cheeses make a tight, time-tested lineup that best utilizes seasonal fluctuations and splits milk into a range of textures and culinary applications.

Pecorino Siciliano is a large-wheel, long-aged cheese that is often grated onto pasta. Sheep milk is coagulated with lamb rennet, no added starter. Many aged pecorinos (sheep cheeses) from Sicily and Sardinia possess a strong picante sharpness, a flavor often originating from rennet pastes high in **lipases**, enzymes that break down fat. This lends these cheeses a stomachy, mouth-numbing, sometimes bitter or acrid flavor that is divisive but highly addictive once a taste for it has been acquired. I know this doesn't sound appetizing, but it's like fish sauce in a stew; it adds a background boost of deliciousness. Parmigiano Reggiano has a slight hint of this lipase character. Pecorino Siciliano has a moderate dose, being one of many flavors. The meaty animality plays a supporting role to the notes common in aged sheep cheeses: tropical fruits, sweet caramel or butterscotch, lamb and lanolin. A spoonful of the rennet paste is measured on a scale, mixed with warm water to make a slurry, then poured into the milk while stirring vigorously.

The milk sets into a firm curd within the typical time frame I see around the world, forty-five to sixty minutes. A wooden tool called a *rotula* (kneecap) is used for cutting. It consists of a slightly bowled disk attached to the end of a long pole similar to a broom handle. Rather than careful and delicate cutting, a violent whisking motion is employed, rapidly breaking the coagulum into small pieces sitting within a milky whey. It shocked me to see this, as my mind was trained to see this as a "loss of yield" since much of the fat had been lost to the whey, resulting in less cheese. While stirring rapidly, hot water is poured in to raise the temperature. The curds firm up with about ten minutes of stirring, and then are allowed to settle to the bottom. The whey resting on top is pumped into a large stainless vat, where it is heated up for ricotta. Once the curd below has come together as a matted mass, it is scooped out and placed into big plastic basket forms on a draining table. The curd hunks are broken up in a motion called *intumare* (to bury) in which you poke at the pieces with pinched fingers in a pecking motion. This breaks apart the ragged curd, allowing more whey to be released and closing up potential pockets inside the cheese.

Once the ricotta is finished, the large pecorinos are soaked, in their forms, in the 160°F (70°C) scotta. They are left for an hour and a half to two hours as the whey cools, inducing a thermophilic fermentation and probably knocking out many microbes. It also helps the surface of the cheese seal up completely. It's an interesting and unorthodox approach. Instead of exposing the curd to high heat in the vat, they do so after forming the wheel, using the heat generated in making ricotta. After this heat treatment, the wheels are given their final, rustic shape in handwoven baskets called *fasceddi* made from a juncus reed. These baskets are a great example of the use of local materials in an ancillary craft that is supported by regional cheesemaking traditions. The DOP requires the use of these forms, which helps support the artisans who still make them, showing that a protected name is a device that can be used in many ways, sometimes actually achieving the goal of supporting traditional tools and practices.

The wheels are brined, then rubbed in salt as they sit on wood shelves in an aging space attached to the caseificio. This pulls moisture out and gets more salt into the cheese. After one month, wheels that have no visual defects can be sold to an affineur, who will handle aging and distribution. By this time, some of the wheels may have begun to swell, an indication of gas bubbles forming inside the cheese. This is likely due to the amount of fecal bacteria in the milk, the amount of time that elapses between milking and cheesemaking, and the absence of a tangible, active starter culture. Without a starter, cheesemaking is always more of a gamble. These slightly swollen cheeses and those with cracks on the rind are kept at the caseificio to be sold as non-DOP pecorino and eaten by the makers. Cheesemakers and their families often don't eat the best cheeses they make. They eat the imperfect, the leftovers, and the ricotta.

The making of pecorino ramps up in the winter when the milk is at its peak in terms of production and potential for aged cheeses that express botanical terroir. Pasture vegetation and hillsides are lush from winter rains, offering sheep in the middle of their lactation a diversity of herbs, forbs, legumes, and flowers. You can't simply make any cheese well, at any time of the year, from any milk. Especially with milk from a herd that is bred seasonally, meaning they all give birth within a certain time frame. This is even more true with sheep milk. Early- and late-lactation milk can be abnormally rich and is not well suited for long-aged cheeses. Too much fat impedes the removal of moisture, and the milk can be microbiologically problematic.

Vastedda

The second cheese is a stringy one made by setting aside some of the pecorino curd, then stretching it. Once any rennet-curd cheese ferments into a certain pH range, it has the property of being able to stretch when heated to around 170–180°F (76–82°C). This allows the curd to be transformed into long, toffee-like strands and then shaped into various formats, seen in the range of sizes and shapes in Italy's pasta filata cheeses such as burrata and scamorza. After the curd is exposed to hot water, it becomes like malleable molding clay and can be given any shape before it cools and becomes solid again. As with the pecorino, the level of heat treatment alters or steers the **microbial ecosystem** of the cheese and can neutralize potentially problematic microbes.

The curd is made into a wheel and left to drain while slowly fermenting in a cool place until it is ready to stretch. In the summer, it ferments faster and is ready in twenty-four hours; in winter, it tends to take forty-eight. The tricky thing with this style is the timing. You have to catch the cheese while it's in the window in which it can stretch. If it ferments too far, becoming so sour the pH drops below 5, it loses this ability. With enough experience, you can tell that the curd is ready by changes in texture and aroma. The texture softens as it ferments and becomes less squeaky. It gives off distinct aromas of milk that is souring but still pretty sweet. I've seen people in Italy simply glance at the curd and say, "It's ready." Most of us have to take a little piece and expose it to hot water to see if it stretches.

I found it perplexing that the cheese was ready at either twenty-four or forty-eight hours, fitting perfectly into the day's schedule. What if it had reached the stretch zone in the middle of the night and by morning was too sour? A slower fermentation gives a wider window in which the curd will stretch, and the maker can speed up or slow down the process by moving the curd between a warm room, cool room, and refrigerator. Food is often elevated when we allow it to take its time or intentionally slow the process down, what sourdough bakers call retarding the fermentation. This opens doors to the possibility of enhanced depth of flavor in bread, and I was beginning to suspect the same was true of some cheeses.

Years of trial and error and the work of many hands will iron out a recipe that doesn't readily translate into words on a page. A technique evolves that requires hands-on training and a sharpening of the senses rather than precise numerical recordkeeping or a background in dairy science. There was no pH meter in the caseificio. There was a photo on the wall of Nono Calogero, and Nona lived in an apartment upstairs, occasionally coming down to stir the vat while smoking cigarettes. This depth of cultural continuity is what makes these cheeses what they are. There is no secret recipe you could learn and then replicate elsewhere. You can make a stretched-curd cheese anywhere in the world, but it won't be the Cangemi's vastedda.

The curd is cut up into thin slabs and placed in a purpose-built, coopered wood vessel called a *piddiaturi* that is wider on the bottom than the top. If the timing works out, the ricotta has just been ladled out from the day's whey and the remaining scotta is still piping hot. It is poured onto the curd, which is slowly stirred and lifted to disperse the heat. A wooden paddle is used to lift the mass of curd as it slowly stretches into a long, elastic sheet, which is brought together as a smooth blob. Small wads are pinched off and placed into flat-bottomed soup bowls that give the cheese its final shape. After a few flips, the perfectly smooth disks are left to cool, salted quickly in brine, then packaged to be consumed within a few weeks. I preferred the cheese at four to seven days when the texture became softer and more cohesive, with a pleasant chewiness to the rindless outer skin. The caseificio makes vastedda to fill demand, with any remaining curd being used for pecorino, which has a shelf life of years rather than weeks. The cheese is one of a kind, with a sheepy profile of lanolin and fruits wrapped up in an eminently snackable puck of fatty string cheese that melts well and can fit in on any cheese plate.

It is somewhat rare to find pure sheep milk stretched-curd cheeses. Curd from sheep's and goat's milk doesn't stretch as well as that of bovines (cows, buffalo, yaks). I've never been able to get sheep milk to stretch the way it did at the Cangemis, where I witnessed what is called infinite stretch, in which the curd can stretch many feet without breaking. Some say this is only possible with the milk of the Belice breed; others that it is an effect of sulla, the legume that is an important part of the sheep's diet. It could be both of these factors, combined with the slow, natural fermentation. Cheese is mysterious, and it comes out better when its complexity is more than we can claim to contain intellectually; when we celebrate it as an enigma, as a partner to dance with rather than a set of parameters to control. Sublime cheese can't come from statistics or from the fine-tuning of dials and measuring devices. Deliciousness and bliss can't be quantified, and they can't be forced into being through monetary investment, recipe development, or a spreadsheet.

I was beginning to realize that these ecstasy-inducing cheeses I sought would come not from a plan in the head, but from hearts that beat and legs that walk. They come from a landscape that is maintained by the livestock that grazes and inhabits it, from the farms and buildings set up within the constraints of that place. They come from people who love and care for the places they call home. As my friend Max Jones put it while speaking eloquently about threatened foodways, "The people are the recipe."

Ricotta

Many people think of ricotta as being a somewhat boring B-list byproduct. Before coming to Italy, I didn't really know what ricotta was. I had never experienced how good it could be. In the United States, what is sold as ricotta is almost always made from whole milk with an acid added to cause the bastardized ricotta to form. What is available in Italy at essentially every small-town caseificio, every time they make cheese, is something else, made from whey, sometimes with a bit of milk added. Italians are connoisseurs of ricotta and line up to buy it steaming hot from the vat.

In Sicily, I had a pinnacle ricotta satori in the sheep milk manna pumped out by the Cangemi caseificio. It was served in a novel way, as a cheesemaker soup. *Zabbina* is made as a snack for the makers, who by midday have probably only had coffee and a pastry for the pitiful debasement that is breakfast in Italy. Lunch isn't until two. So as the morning's work is winding down at eleven or twelve, ricotta is ladled into a bowl with some of its hot scotta, which is surprisingly rich, with a cooked-milk flavor amplified by heavy salting. Day-old bread is crumbled in, soaking up what amounts to a broth. The mixture of soggy bread, hot whey broth, and delicate ricotta is slurped down noisily while standing, a quick break before getting on with the janitorial work that is often a major part of cheesemaking.

Words cannot convey the beauty of this ricotta, but let us see how close they can get. Satin pillows cased with a silky sheen are freed from a steaming lump of milk love by a spoon that now will never want to be dipped into anything else. Slid between one's lips and onto the tongue, is the texture slightly jelly-like or is it more of an airy soufflé, bursting with pockets of hot whey? There is a distinctly eggy flavor and mouthfeel reminiscent of custard. In zabbina, this takes on an egg drop soup aura. It's less creamy or fatty, and more a protein-rich concentration of milk's umami side, alchemized by heat and salt. All this is tied together by the milky sweetness of lactose elevated rather than fermented. It all melts like butter in your mouth . . . an ethereal, transitory moment of bliss. Followed by the next spoonful. Life, like ricotta, is fleeting. That's what makes it beautiful.

How is it so good? I have asked myself this many times. It is so good because anything less would be unacceptable. Sicilians know how good it can be and perhaps have fond memories of mothers and grandmothers making it. It is a part of their food culture, used extensively in a fleet of sweets, in which cannoli is just one vehicle. One factor is the violent cutting of the coagulum, leading to a richer whey and therefore a richer ricotta. Second is the addition of a proper dose of salt to the whey, which will then be evenly dispersed throughout the ricotta. The third factor is the little plastic baskets the ricotta is ladled into, which it will be sold in, often while it is still draining. These have tiny slits on the bottom and sides, holding the delicate cheese in while allowing whey to seep out. The pillowy texture would be lost if you removed it from the baskets, to mix in salt or package it in a different container. By not disturbing it, and serving it directly from its draining form, you maintain the frail soufflé texture.

I realized there was an inversion of priorities here. The ricotta was primary, rather than being a byproduct of cheesemaking. The other cheeses were almost byproducts of ricotta making, something you needed to make to get the sweet whey. From this perspective, the inefficient cut leading to lower yields of cheese actually led to more, richer ricotta. The experience of unlearning that had begun in Mongolia continued as I traveled and saw every supposed rule of raising livestock and making cheese broken or inverted.

The ricotta of Sicily taught me in a language we can all understand. A language made not of words but of delicious bites, enticing aromas, unnamable textures, and the guttural utterances they evoke from our throats as we express the spontaneous stirrings in our hearts. A language of waggling sheep tails, bells jangling, fingers brushing the bark of olive trees, bare feet crushing fallen orange leaves, wheat stalks waving in the breeze, composting manure generating heat with mushrooms growing from its sides like trees on a volcano. A shepherd calls out and waves a staff while seated in the shade of an almond tree. The sheep turn to stare, then resume nibbling, the rhythm of their grinding teeth soothing the earth as their hooves massage her scalp.

— CHAPTER EIGHT —

A Contract with Goats

KELMEND, ALBANIA

The people cannot be healed unless the land is healed. The land cannot be healed until the people are healed, by the land.

—MARTÍN PRECHTEL

On a sunny April afternoon, I sat outside a stone house in the Accursed Mountains of Kelmend, northern Albania. Thick blankets of snow kept the craggy peaks cozy as they slept; the valley below was just emerging from hibernation. Spring was singing, the creek flowing hard with a merry song of humor and unassailable optimism. I had come here searching for a cheese called Mishavinë that is aged inside coopered wooden barrels. Hefty shards of it sat on the table in front of me, which was slowly filling with an excessive amount of food: thin slices of cured ham, stacks of home-baked bread, the feta-like *djathë i bardhë* (white cheese). I picked up a big hunk of crumbly, milled-curd Mishavinë, inhaled deeply, and took a nibble. My eyebrows lifted like alarmed caterpillars. I can normally start rattling off tasting notes and comparisons, but this cheese had me stumped, in the best way. I took a real bite while gazing at the peaks that embraced the sheltered valley. I heard the birds chirp, and nodded slowly, started breathing again. Everything was in its place. This is the bliss-of-pure-awareness reaction I look for, cheese satori. No words . . . my mind scrambles to find an edge of familiarity, then lets go as I enjoy simply being alive, a new food and place beginning to course through my veins as powerful medicine. The old man we had come to see smiled at me, appreciating my reaction from a place inside himself where language barriers don't exist. We all ate and drank, feeling the sun on our flesh as the snow melted and ran down the rocks on its way home.

I had arrived in the seasonally occupied village of Lëpushë with my local guide, Fran Lumaj, who would drive me around Kelmend in his Mercedes for the next week, weaving around road cattle and flocks of surly sheep, crossing rivers, crawling up mountains, visiting shepherd families and remote villages that felt like ghost towns, with some hangers-on who were too weird, poor, or proud to leave. People are fleeing the depressed region for the city or to emigrate and work abroad. A degree of food sovereignty survives in Kelmend, the remnant of a mountain shepherd polyculture of dairy animals, hogs, bees, tree crops, foraged foods, and medicinal plants. Unfortunately, there is now a negative connotation to practical shepherd foods, with many Albanians seeing them as symbols of a past they want to move beyond, of extreme poverty and the Marxist dictatorship that didn't formally end until 1991. In a country where these imported items were not available until quite recently, Coca-Cola and brightly colored bags of chips or candy are symbols of freedom and progress.

Fran took me to the guesthouse where I would spend the next week. My room was an out-of-place new addition that teetered arrogantly off the edge of a rock outcropping with an eagle view of the whole valley. It was contrasted starkly by the old stone house where the Dragu family has lived for generations, tucked sensibly on a flat spot on the hill, shaded by tall trees. We met Tom and Luçie Dragu. Tom is a husky man with a bull neck and contagious smile. Luçie has long, wavy black hair and a quiet, hardworking, motherly demeanor. Luçie took us to see the pantry where she stored her white cheese in tall plastic barrels of brine made from post-ricotta whey. Boards and rocks placed on top of the cheese kept it submerged while wispy kahm yeast and white mold rafts swirled around on the surface and coated the rocks, like seaweed on a rocky shore at low tide. Next to this were bins full of pickled green tomatoes, peppers, and sauerkraut. Hams hung from the ceiling. She didn't have any Mishavinë to share, but agreed to demonstrate how it was made.

The morning's cow's milk was heated and powdered FPC rennet added. Cheesemakers in Kelmend made the switch to this manufactured stuff in the 1990s and see it as progress to not have to continue making their own rennet, from their own herds. To buy imported rennet is modern, a step forward. Many people told me that nobody was making their own anymore. Fran and I later randomly met some elderly women watching their goats, and when I asked (as I always do) what kind of rennet they use, they whipped out a gallon jug with goat abomasa soaked in post-ricotta whey, and a few holly leaves, which are also left as offerings on shrines. It was an important lesson: Even when experts and locals tell you something has disappeared, it may still be there somewhere. Perhaps they haven't looked or asked that question. Maybe they don't want to see it, or don't want you to. Besides, the old people in the hills have a way of keeping secrets, and may not be sharing the names of all the seeds in their shed. The young white cheese from these ladies was exceptionally good, made from the milk of their long-haired goats who munched the dried leaves off an alder branch that had been cut and stored as tree hay.

We waited until the coagulum was firm, then Luçie cut it with the handle of a soup ladle, first making the sign of the cross. I had seen this before. In Catholic regions, it was always described as a Jesus-referenced blessing for the cheese, to prevent evil influences. The people of Kelmend are proudly Catholic, a part of their cultural resistance to the Turks, who halfheartedly forced most of Albania to become Muslim in the fifteenth century. Albania today is marked by religious plurality, with people commonly marrying those of other religions and attending more than one church. Catholicism in Kelmend feels like a veneer over much older forms of worship, the nature-centered animistic beliefs that are despised by most monotheistic religions. It feels like Kelmend was never successfully conquered by any imperial army or inquisition. The cross is not just a Christian symbol; it is much older. It can be associated with the idea of the "world tree," which shamans climb up into the heavens, or down into the underworld, to do business for those of us on the surface. I kept seeing blood-red crosses painted on the doors of livestock housing, with added lines on top and at the bottom that made it resemble a mushroom. Combined with a ram skull over the doorway, the scene looked more like the cover of a Gothic black-metal album than something out of Sunday school.

Luçie finished cutting the curds by whisking them until they were quite small. She scooped them into a cheesecloth and pressed them into a round between two plates with a rock on top. She made *gjizë* (ricotta) and explained that, after the small wheel stopped dripping whey, it would be left to dry outside, hung in a little grilling box. After five days, she took the wheel, cut it into strips, and fed them through a meat grinder forming pasty strings. This was salted to taste, reshaped into a fat patty, and packed into the bottom of a tiny wood vessel. Luçie said this demonstration was just for me, as the cheese isn't made in April and typically involves the milling of multiple days' worth of curd, packed in a three-gallon wood vessel.

We learned more about Mishavinë by visiting a farm with a guesthouse down the road. Fran is a schoolteacher and one of his students lived here, a sixteen-year-old boy named Noja who, in addition to being a shepherd, was an aspiring mountain guide. Like Fran, he said he hoped that tourism would bring in enough cash flow to allow him to stay, and his family to keep farming. The family at Guesthouse Trojani exhibited the default hospitality of Albania, and of many communities whose cultural norms based on innate human kindness haven't been totally swallowed up by the leviathan of consumer culture. Come into our home, sit down, drink coffee and *rakia*. Piles of bread, bowls of yogurt, sliced ham, white cheese, and Mishavinë. I still could not describe its flavor. To truly get to know the cheese, I would need to sit down with a larger sample size—a big chunk to eat slowly over hours, with different beverages and snacks. The most striking immediate impression was textural: The cheese was a mash-up of shards, crumbles, and creamy pockets, resembling a geologic conglomerate of river stones and sand, pressed by time.

Noja explained that his family made Mishavinë in August, September, and October with full-fat sheep milk. In the scant literature about it, there is talk of it being a mixed-milk cheese, combining cow with sheep and/or goat milk. It turns out it is a member of a now rare style made by combining small wheels from multiple days into single large cheese, after they are all ground up, mixed together, and salted. There was no starter culture used; only rennet was added. Fresh wheels made daily were pressed for three days before being lightly salted. After drying a further five to ten days, the cheeses from multiple days would be broken up by knife and hand. The milled curd was salted to taste and packed into the wooden vessel. A wood lid was put on and it was left for three weeks, releasing excess moisture through the porous wood. Any mold that grew on top was scraped off and the exposed surface coated thickly with clarified butter, forming a cap. In December, the oldest batches would be pulled out of the pantry cellar. The butter cap would be scraped off and discarded, and the cheese scooped out on demand, to be eaten through the winter.

When cheeses are aged as rinded wheels, their skin is exposed and care must be taken to maintain humidity in their aging space. They are innocent, naked milk children that everyone wants to eat, from microbes to insects, rodents to primates. With Mishavinë, the coopered barrel becomes a little wooden cheese cave, a protective barrier that can absorb moisture while preventing excessive dehydration and allowing some air exchange. The cheese does get some yeast and mold growth where it is in contact with the wood, little veins of white and blue mold, or red and orange coloration and dehydration on this modified rind edge zone. I always looked for those rind pieces on the plate and ate them, finding many cheeses in one. Little pockets of decadence or disgust where life had found a way in, these veins are referred to as intrusions and are often considered undesirable. Many people discard these colorful bites out of fear and hatred of mold. Others consider them the best part. Cheeses like Mishavinë show that the boundary between inside and out is never final. No box can be fully sealed; the rind of Mishavinë is a fuzzy edge fortified by wood, a living transitional zone. A porous border. The edge is where the deliciousness is.

After we saw the family's gorgeous and confident black-faced sheep, their chickens laying eggs inside the calf barn, and the fruit trees that provided for their jams and alcohol, I asked Noja if he thought Mishavinë was in danger of going extinct. He stood remarkably tall for such a small young man and spoke with a composed air of honor and pride while looking straight at me: "The people of Lëpushë will always make Mishavinë."

Kelmendi Cuisine

Back at the Dragu guesthouse, I crept into the silent main hall. An open fireplace threw some heat into the cold fortress of a house. A place had been set for one at a huge banquet table. Luçie brought me wine before surrounding me with plates of stewed pork, fried potatoes, sauerkraut, white cheese, and a tower of bread. The food was simple, high fat and protein, presented with zero frills or spices, the colors quite drab. Farm-to-table looks quite different in Albania: Take the food from the farm, put salt on it, leave it somewhere cold until you're hungry, put it on the table. Every meal would be like this, mainly consisting of meat, eggs, and dairy they had either raised themselves or purchased from neighbors, their own preserved veggies and jams, local potatoes, supplemented with mass-market staples like white flour, sugar, seed oils, and flaccid produce. I would show up hungry to have Luçie silently and without smiling place enough food for three hungry people in front of me, before floating back into her kitchen, not to be seen again as I smacked my lips, laughed at my wine buzz that the fire kept stoking, and tried to get fat in the cavernous hall of a family whose roots anchor that well-settled stone home to the hillside like a tooth in the mouth of the earth.

The simultaneous serving of foods from diametrically opposed approaches to feeding hungry humans would be displayed at every home we stopped at. After rakia distilled from a range of local fruits, a bunch of canned beverages were placed on the table in front of me: sugary sodas, energy drinks, and imported beer. They were surrounded by the artifacts of pastoral food sovereignty. Cheese made by families who milked animals, which they also ate, accompanied by salty ham from hogs raised in conjunction with the dairy herds, home-baked bread, and vegetables grown in gardens then fermented in bulk. It's hard to imagine better food, and much of it reminded me of the alternative local food systems people are working to build or revive around the world, where the fruits of industrial progress have been found to be rotten. Those fruits have made it to Kelmend and pretty much every region I visit with strong pastoral traditions, and they are generally celebrated as something new, cheap, and exciting. Cooking with lard or butter has largely been replaced by unrestrained use of cheap seed oils.

When a food system first industrializes, there is generally optimism. Within a generation or two, there is often a countercultural backlash as people realize how unhealthy these foods are and seek to resurrect the farming, cooking, and preservation skills their grandparents had so they can reclaim the abandoned foodways. As a country like Albania runs as fast as it can toward canned soda and strip malls, people from places like Germany, France, the United Kingdom, and the United States are trying to run back in the opposite direction, seeking places like Kelmend where the vestiges of older ways still exist, where local food systems have been kept remarkably intact until quite recently. I was intrigued by the existence of these older foodways that felt so vital and relevant, while the people I was visiting were often seeking the lifestyle I was running away from. Each of us was likely romanticizing the life of the other, lured by the seductive grass, which is always greener on the other side of the world.

Looking back on my travels, I appreciate the old people who were not trying to go anywhere, who were happy to sit in the shade and watch their grandchildren play while sipping homemade liquor and gossiping with their friends. They could see the value of the old foods and lamented their loss, but kept telling their stories, smiling, and laughing. Laughter and chit-chat are the highest forms of wisdom, and if we are to make it through this tumultuous time, we must remember to keep smiling together as we grieve what is being lost. Without a sense of humor and lighthearted optimism, we are already defeated.

When I asked for *caj mali* (mountain tea) instead of the canned drinks, my hosts were usually surprised, and sometimes impressed. Mountain tea is a name used from Greece to Spain, and the drink is often made with *Sideritis raeseri*, a wild Mediterranean herb. Caj mali is the tip of an iceberg. The pastoral lifeway is marked by a knowledge of healing plants, where to harvest them, and how they can be used to treat both humans and livestock. Herders are not just keepers of livestock; they are foragers, hunters, fishers, naturalists, herbalists, veterinarians, meteorologists, mechanics, storytellers, and musicians. While in Kelmend, I spent time with a herder named Fatmir, who sported the classic style of an older Albanian man: leather shoes, dark slacks, a collared button-up shirt under a sweater and blazer. He wore this while he led me on a hike to some abandoned summer grazing settlements, where he picked medicinal plants and we sipped from a bottle of rakia he kept stashed next to a spring. He was remarkably fit for a man in his late sixties, and set a pace that I could barely match. A life spent chasing goats around the mountains, eating organic food, and making medicine will do that. He shared with me that, according to their legends, livestock taught humans the traits of various healing plants and where mineral salt deposits and springs were located. Specifically, goats, the only farm animals intelligent enough to be anarchists.

Milk as Medicine

A seasoned, observant herder watching over an unfenced herd possesses an intimate awareness of the health status of the collective and its individual members. If an animal expresses certain symptoms, the herder keeps a close eye on its behavior. How much is he drinking? How does she interact with the herd and vice versa? What do his eyes, mucus, and poop look like? What is she eating? Goats are renowned and despised for feeding on a wide range of plants. Released in a field, they often run straight through the grass to the trees, hedges, or scrubby margins. They seem to prefer the woody, spiny, and bitter plants containing medicinal compounds that at one dosage may be toxic, at another beneficial. Sometimes the poison is the cure. Observing that the sick animals focus on eating particular plants, herders realize they are self-medicating, and take detailed mental notes. When you see a rancher, herder, shepherd, or farmer standing at the fence silently and seriously watching their animals, they are engaging in an exercise humans have engaged in for thousands of years.

Such observations have accumulated over generations as herders have experimented with the medicines on their herds and themselves, slowly building up a pastoral pharmacology. While the herd is feeding or ruminating, herders perform other tasks such as spinning yarn, whittling implements, cutting wood, foraging, and collecting medicines. Herds are taken to highlands every summer not just so they can graze, but also to replenish stores of medicine, berries, and mushrooms. These herds consisting of old breeds have been traversing these lands for generations, exercising what behavioral ecologist Fred Provenza calls their "nutritional wisdom." Like bears, they know where the tasty and healing plants are, and when they are in season. Fatmir stated all this in passionate oration, which was translated for me by a French anthropologist: "We didn't domesticate goats; they invited us into their herds. They learned our language and taught us theirs. They offered us a contract: They give their milk, flesh, and hair in exchange for shelter from the elements with beds of straw, food for the winter, and protection from the wolves." It is a symbiotic, reciprocal relationship. "But now people just take as much as they can, then push the animals and the land to produce more, without giving anything back. The animals and the land miss us, they long for our songs, music, and footsteps. We've broken the contract, and turned family members into slaves."

Fatmir was adamant about honoring the strict social norms codified in the Kanun, a set of tribal laws famous to the outside world for the phenomenon of "blood feuds." He said these include a dietary component, which dictates that each milk is a gift from God that should be kept pure. The so-called shepherds that were mixing milk had turned their backs on the laws, and were illegitimate. To mix the milk of different species is a blasphemy against God. While I admired Fatmir's passion, it was becoming clear that there was a high degree of factionalism in the herder community, and a lack of agreement over how to define Mishavinë. This is no surprise; it is perhaps impossible to define a "traditional food." Who gets to define tradition? Is some essence of it lost in the attempt? *Tradition* is a dangerous word. To name something is to limit it.

The next day, I heard a story that carried forward this idea of milk as medicine. The French anthropologist I mentioned above came to live in Albania and work with the herders. She arrived in nearly constant pain from a stomach condition that no doctor had been able to diagnose or treat. She mentioned this to a group of herders, who began to ask her very detailed questions. Then they said, "Come back in a week." A week later, she came back. They gave her a wheel of fresh cheese, made from goat milk. She ate it over the course of a few days. Her stomach condition disappeared, and has not returned since.

The herders told her what they had done. They took the goats to specific plants that, according to their pastoral pharmacology, would be appropriate for her ailment. After encouraging the goats to remain on these plants for the afternoon, they brought them home and milked them. The milk was used to make a single, medicinal cheese. For a long time, I interpreted the story like this: The plants contained healing compounds that transferred into the milk, and then into the cheese. Recently I heard another interpretation from an herbalist, who believed the herders had influenced her gut microbiome by first altering the goat milk microbiota on that day, via bacteria on the plants, or how plant terpenes played a selective role in the preexisting milk microbiota. They then cultivated that altered milk microbiota via fermentation into cheese that introduced a thriving pouch of the right seeds into her digestive tract, correcting an imbalance.

Vermosh and White Cheese

Just before I left Kelmend, Fran took me to his village of Vermosh, close to the border with Montenegro, to introduce me to his family. I later realized I should have done this on arrival, requesting to meet the "chief" of every village I had visited, to ask permission to be there, poking around looking for stories and recipes to extract. In my mind's theatrical version of this phase of my journey, I am John Wayne strutting into a bar, looking for cheese recipes and photo ops with quaint shepherds rather than whiskey and fights. I'd like to say my intentions were pure, but behind it all there was a vague aura of the culture I was fleeing from, and its mindset of conquering, of extracting the treasures of other lands and leaving, hauling the riches home to show everyone how interesting my life was. That may be an exaggeration, but it's a point I feel needs to be made, as I recognize that the discussions around cultural preservation and the protection of tradition can be a tricky ground to walk, covered in good intentions that can go horribly wrong.

Fran's father had passed away a month before, which explained some of his heaviness, as he was now the head of the household. Fran and I sat side by side at a table, as plate after plate appeared from the kitchen. Crispy fried trout, sliced potatoes cooked in cream, sauerkraut, Mishavinë, and white cheese. This was the first time I was served *kacimak*, a corn polenta enriched with butter and cheese that is a staple food across the Balkans. The meal was phenomenal, and I washed it down with glass after glass of the house wine that they had huge barrels of in the cellar. Some would call it vinegar, and there were specks of broken-up kahm yeast pellicle floating in it, about as natural as it gets.

I asked to see the cellar, which I realize is somewhat of an odd request in many places. I'm not sure if this is due to embarrassment of the old peasant foods that have negative connotations, or if it's like asking to see where someone keeps their money. The cellar was a half story underground, kept cold by thick stone walls that held in the night, extending winter's chill. Sacks of potatoes, walnuts, and apples, canning and winemaking implements, jars of red peppers, eggplants, pickled garlic. I was most interested in seeing what is called white cheese around the Balkans. It is essentially feta, but if you are in Albania, do *not* call it that. It's kind of a sensitive issue, due to long-standing political and cultural tensions between Greece and its neighbors. But I use the term *ricotta* generically for that style, and I use *feta* for these brine-aged cheeses that are made across a wide geographic area, since there is a widespread familiarity with it. These foods transcend nation-state boundaries, but many are caught up in **gastronationalism**, with various communities, ethnic groups, and countries claiming them as their own.

Feta-style cheeses are frequently made from sheep and goat milk, and often have a high level of acidity. They can have a variety of textures but are unified in being stored long-term in salt water, ideally kept submerged. This is a strategy for aging cheese in hot, arid climates, using salt to keep the cheese in a kind of stasis. I'm not sure we should consider it an aged cheese, even though Fran told me they sometimes store it for over a year. It's more of a pickled cheese. The acidity and anaerobic storage in a highly saline liquid slows down the chain reactions that normally occur as cheeses age through fat and protein breakdown, and rind microbes doing their work. That's kind of the point, using salt to put the brakes on heavily, to slow down biology's unremitting lust for composting itself.

In my opinion, brine tanks require skimming to remove rafts of cheese goop on the surface, which molds can stitch together as islands that begin to grow forests of filamentous fungi. Even with no goop, without cellar temperatures, you will likely get salt-tolerant yeasts forming colonies on the surface, providing a space for life to do what it does . . . grow. I age feta in vessels with zero airspace, or I stir the surface, flip the jar, do something to break up the surface yeast formations once or twice a week. I'm into many moldy cheeses, but this is one style I feel does not benefit from rampant growth. I'm not looking for funky feta; I want to taste salty cream, bright acidity, and a bit of animality. Most of the white cheese in Kelmend that I tasted were kind of gnarly, besides some that I suspected were from industrial dairies, or very young stuff. The people here say they love it, though; for them, these are the flavors of home, true to style.

The brine tanks I saw in Kelmend had crazy furry pellicles of wispy kahm yeast, wrinkly *Geo*, blotches of blue and gray. Different homes had different brine pellicle ecologies, due to variations in salt percentage, acidity, and temperature. When they revealed these brine pellicles by pulling back the pillowcase held by a large rubber band, it was obviously considered completely normal to them. Rocks were often used to keep the cheese submerged, but the brine itself takes on the fungal flavors. Brined-aged cheeses can only taste as good as their brine. These are of course highly personal judgments. With respect for cultural relativity, I have strong reactions to food and see value in voicing them openly. I believe in being honest about how cheeses make me feel, and letting these reactions inspire my words.

There was a diversity of flavors in the white cheese from various houses, but almost all of it was what most of us would call oversalted. Too salty to just eat on its own. Whatever flavor was in the milk, whatever potential for botanical terroir it contained, is overwhelmed by too much salt. I think this moldy brine and heavy salt have a type of terroir, considering these characteristics belong to a cheese that has been part of a pastoral cuisine over a long period of time. Terroir doesn't necessarily taste good to everyone.

My time in Kelmend left me with conflicted emotions, and little clarity. I was here outside the milking season and didn't get to taste many of the foods. It was one of the few places I've visited where I could feel that some people were skeptical of my intentions, and that there was a degree of reluctance in sharing the information I was seeking. Which is understandable, and fits in with the history of the region. The remarkably intact food systems, old breeds, and deep folk knowledge are still here because armed tribal groups have stuck to rigid social norms and resisted incursions of outsiders. These are highly conservative, guarded, insular people.

In what may at first seem contradictory, one aspect of these old customs is that guests are to be honored and showered with hospitality. I suppose it's a case of keeping a watch over who is coming and going, guarding the passes, but also opening your doors to travelers who are obviously not a threat. You fight off invaders but welcome the hungry. You don't maintain all these stores of food and alcohol so you can lock yourself in with your family. You have it to offer your guests, in lavish displays, as if making offerings at an altar. Even if some people here had understandable apprehension about my intentions, they didn't hesitate to welcome me in and feed me, making me comfortable as a matter of course, following cultural protocol. The food in our pantry is not really ours. It's a perpetual blessing, and we need to feed all life—the birds, the bees, and the soil like they are children or livestock. Life feeds on life, it feeds us, and we must feed it. That's the agreement, and a cheese like Mishavinë is a relic of this ancient contract.

— CHAPTER NINE —

Sheep Hides and Salt

TUSHETI, GEORGIA

Shepherds know many mysterious languages; they speak the language of sheep and dogs, language of stars and skies, flowers and herbs.

—MEHMET MURAT ILDAN

I've become fascinated with many cheeses over the years. Certain cheeses pull me in like a bee to a flower, and the less this draw is based on logic, reason, or gastronomic sense, the stronger is its force. The best example of this is a cheese called Tushuri guda that is made in the former Soviet republic of Georgia. Georgia is a tiny country that sits sandwiched between Turkey and Russia, with the Black Sea on its western flank. The region of Tusheti is in the Caucasus mountains right against the Russian border, a zone of great linguistic and ethnic diversity, and ongoing disputes between imperial powers. In the summer months, herders bring flocks of sheep, along with cows, goats, and horses, from the Georgian lowlands up into these mountains, a vertical transhumance they have been practicing for hundreds of years. Very few people live here year-round, but many families come up in the summer to stay in villages, often clustered around stone defensive towers built in the seventeenth and eighteenth centuries. They are increasingly building guesthouses or converting their homes for tourists who come to ride horses, trek between villages, and enjoy the pastoral culture of the Tushetian people in a land that remains fairly isolated. I heard the joke more than once: "We used to milk sheep. Now we milk tourists."

The famous cheese of Tusheti is an orphan from a family whose origins stretch back beyond our sight, into the mists of history. Tushuri guda is aged in a sheep hide turned inside out, so the cheese is in contact with the woolly side. Recently, the sheep hides have been replaced by plastic bags. Referred to as cheese-in-a-sack or skin-sack cheeses, these relics fascinate me as emblems of the reliance on animals that is the keystone of pastoral dairying. Livestock can provide not only the milk, but also the rennet to turn it into cheese and the hide that the cheese will be aged in. To take the milk from the animal, perform the biological process of cheesemaking outside its body, and then put the cheese back into the hide, which fills up and becomes a rough sheep shape again, is astonishing in its purity. Cheese, meat, and hide all intertwined in the perfect symbol for preindustrial approaches to working with milk, living with livestock, and living in trust rather than fear of microbes, of life.

The cheese-in-a-sack family represents the opposite of cheeses made in a hospital-like sanitized room full of stainless steel, from milk with a compromised immune system, aged in a temperature-controlled warehouse. If a cheese like guda—made on wood tables set on dirt floors in a shepherd camp, aged in a sheepskin—can be safe and trustworthy, then how many of modern food safety requirements are actually necessary? It's an extreme example that makes the middle ground of farm-made rennet, natural starter cultures, and using wood implements appear mild by comparison. Of course, the stainless-steel-and-sanitizer mentality and corresponding lab-made cultures and rennet are creeping into this remote region, like everywhere on the planet, as the inquisition tolerates no alternative approaches and fears the example set by people who can feed themselves, make their own intoxicants, collect their own medicine. People told me no one was using the sheepskins anymore, that this a thing of the past. Still, I was driven to go, to walk the trails and see the herds, meet the shepherds, and ask them in person. To see if maybe, in some harder-to-reach camp, a bit farther up the trail, some stubborn, cranky eccentric was keeping this practice alive and not talking about it. The past has a funny way of sticking around, hiding in forgotten corners, then popping into the party uninvited, chin held high with an out-of-place dignity and handmade clothing, embarrassing modernity's claims of progress and triumph. No cheese has pulled me in so hard; there was no way I could resist the voice echoing out from the rugged mountains, a consoling lament speaking of ancient rituals living on in inaccessible valleys, in the liminal space between earth and sky, near the crooked and exposed backbone of the planet.

The voice felt like the warm body of a mother sheep,
Whose wool I longed to push into, snuggle up in,
To warm my lamb soul, nurture it to life through love,
After a winter in modernity's cold, spiritless universe of accidents.
I needed to warm my bones next to her heart fire.

Georgian Wine

We can't talk about Georgia and not discuss wine and hospitality. Hospitality in Georgia borders on hostage taking. You can leave tomorrow, they'll say. Today we drink. Then tomorrow it's the same story. It almost felt like bad form to stay only one night. What's the rush? Sit down. Eat. Drink. Stay warm. Wine in Georgia shatters the neat categories of white, red, orange, and rosé. A whole rainbow of colors exists between these, and you can taste your way through the realms of possibility in natural wine. From the clouds of heaven to the cellar floor. Every "wine defect" is on display and consumed without hesitation. Mousy, ropy, and sulfurous are common; Brettanomyces and kahm yeasts run wild. Some of it is exceptionally good, especially the bottles from a new generation of natural winemakers who are pushing boundaries. Home winemaking is ever present; many households have a vineyard plot where they harvest every year, crushing and fermenting in their basement, often fermenting with extended skin contact in giant pottery vessels known as *qvevri*. Adding yeast or conditioners is rare, as eight thousand years of practice have informed Georgians that the wine is already in the grapes, waiting to be released. The fruit wants to be wine, the way milk wants to be cheese. The custom is to have an unlimited supply of both to offer your guests.

I sat in the old stone house of a well-known Tushetian winemaker, brewer, chef, and farmer named Shota Lagazidze. After brief introductions, we began tasting, and Shota jumped into theatrical toasts, setting the heartfelt, emotionally charged mood that infuses Georgian rituals surrounding drinking. We shared our philosophies on life, expanding on our areas of overlap as we drank many bottles, living as brothers for a few hours. When I asked him why he only charged the equivalent of ten dollars for his exceptional bottles, he said, "I don't need more money. I sell enough wine to live the life I want, and most of the wine I make, I drink. I want to spend my time living, grilling meat and drinking with my friends and family. We can sell our time as labor, but we can't buy more. So I don't sell my time. I savor it. I live my life. It's not for sale." I chewed on that wisdom while savoring the naked, unprocessed truth of the wine, which conveyed the essence of his words in liquid form.

At the perfect moment, Shota brought out the cheese. Tushuri guda, the cheese of his people, my biggest cheese crush. I blushed and looked at the ceiling when it entered the room. Shota left to attend to other guests and I sat alone, getting to know the cheese, crushing it in between my fingers, nearly sticking it up my nose as I snuffled and snorted like a perplexed old bloodhound, searching for the trail of a strange, unknown critter. I knew this was a sheep cheese, made in the summer months, aged in plastic bags with a heavy dose of salt. It was over one year old. I held up a tan-blond chunk and noted large gas bubbles that would be a cause for concern in many countries, an indication of fermentation gone wrong, or "contamination by fecal bacteria." It smelled of spicy, whole-stomach rennet, with the distinctive aromas of fats broken down by the lipase enzymes this style of rennet is known for. Stomachy with note of organ meats, animal, mustard, campfire, and adventure. It tasted like a sheep barn that had been left unmucked all winter, minus the ammonia. Sheep poop and pee, wet wool, and a nearly overwhelming dose of salt. The surging vibe of a cheese satori saturated my limbs with a sedating numbness as I sighed, paralyzed in my chair, afraid I would start drooling.

I shook with an insight, that the flavor was primal, what cheese may have tasted like hundreds of years ago, before flavors were cleaned up, when cheese probably tasted like what and where it came from: animals, sheep hides, mountain pastures, salty mineral springs, muddy trails. It wasn't delicious, but it gave me goosebumps. It challenged my twenty-first-century palate, born from tame cheese counters and food stripped of its voice. This cheese is beyond rugged. Tushuri guda is as wild as a coyote laughing maniacally in the moonlight. Most cheeses are fairly benign things that we consume without much apprehension. This cheese was kind of scary. It had a mind of its own and almost seemed to be consuming me.

There are cheeses you bite,
And there are cheeses that bite you.

Shota explained how to get to Tusheti from his village. Every morning, drivers would hang around the main square in small groups, smoking cigarettes, shooting the breeze, looking out for potential customers. Once their lifted four-wheel-drive vans were full of passengers and cargo, they would expertly maneuver along the rugged, infamously narrow and precipitous dirt road that snaked up to Abano Pass before dropping down to the main village in Tusheti, a journey of four to five organ-mashing hours. Reaching the pass, we crossed a threshold, marked by a tiny Orthodox church. A short walk away was an older shrine, covered in quartz, animal skulls, burnt candles, and coins.

Tushetians have dual cultural identities. In the lowlands, they follow the standards that the Georgian Orthodox Church is increasingly pushing as a national identity on a country of many tribes, languages, and mythologies. They eat pork and worship the saints. On the other side of this pass, some of the older, animistic ways are resumed. Pork is banned, and the worship of rivers, mountains, and sacred brewing vessels is practiced, though with crosses imposed on top of the shrines, and small paintings of saints placed next to moss-covered sheep horns. Tusheti is technically a part of Georgia but feels forcibly attached, a fluid nation on the periphery of a state. Tusheti is an in-between boundary zone, once connected to adjacent fragments of a multitude of tribal identities, now separated by a heavily militarized and contested border that was formerly porous. And yet, on the perimeters, in the hard-to-reach, isolated pockets, the state has less of a presence. The old myths and spirits feel a bit more alive in these cultural reserves, which are also often home to the cheeses I seek.

The edge is where the action is.

Sheep Camp

I had a lead, but it's never a sure thing when you are dealing with mobile shepherds. It's not like dealing with people who have gardens and stable Wi-Fi. They usually don't set a date for when they will move up into the mountains or come back down. It depends on the weather, the herd, market pressures, family matters. The plans are generally vague and require a heavy dose of trust in the journey. With hiking poles clicking away, I walked at a feverish pace along a dirt road that wound through abandoned villages, with daunting views down to a river whose growl could be heard miles away. I thumbed every rare car, and got offered a ride on horseback, which I took for sheer novelty. Eventually I reached the end of the road, where the village of Jvarboseli rests peacefully between steep ridges, its stone towers and old homes with ornate woodwork looking as if they had sprung from the ground as naturally as mushrooms. I had been told to come here by a German man I met in Tbilisi who was exploring tourism and other approaches toward creating economic support for the shepherds. He had teamed with a Tushetian man named Mirza, who had a large flock that he brought to a camp farther up this valley every summer. I had scribbled the necessary names in my notebook, which I consulted again for his directions: "Go to the end of the road. Then keep going. They'll be expecting you. Beware of the dogs. They are many, and mean."

After spending a night at a guesthouse, I started walking again, with a few days' worth of food in my well-worn backpack. I followed a trace of tire tracks crossing and recrossing a stony creek, weaving their way up the valley, which eventually opened into a stunning grassland nestled between steep ridges. I could see tall stone defense towers in the distance, surrounded by sheep-mowed pastures, with a wall of tall peaks lending a sense of enclosure, of being held safe in the earth's bony arms. I was entering a forgotten, mythological valley, out of sync with time and the busy world down in the lowlands. This is a romantic trope that I usually resist, but I really felt it in that place; it was a refuge not only of the past, but of an older breed of time.

A shepherd camp was tucked in by the creek, but no sheep could be seen. I walked closer, then cupped my hands next to my mouth and began to holler "*Gamarjoba!*"—the greeting of Georgians, which means "victory." I intentionally did this while still a quarter mile away, as my racket raised a pack of guard dogs from their slumber, causing them to run toward me, barking fiercely, hackles raised. As they did, a shepherd emerged from a stone shack and whistled at the dogs. They continued to bark but calmed down a bit, losing steam as they slowed to a trot. Whining and growling in protest, they begrudgingly allowed me to approach and cross the creek where the shepherd had walked down to point out a small bridge made of logs. "Is this the camp of Mirza?" I asked. "Yes, yes, come on inside and eat," I was directed by universal hand signals from the tall, bearded man. He and two other shepherds in their thirties shook my hand, amused by my sudden arrival. The dogs sniffed me, still skeptical but somewhat accepting, since the herders had allowed my entrance onto their turf.

The shack was built of stones that had been reclaimed from one of the decomposing towers and homes that dotted the valley, mementos of the times when these valleys were inhabited year-round. Blue tarps formed a roof over a birch-pole frame resting on stone walls. A grizzled man in his sixties who had obviously just woken from a nap greeted me with warm handshakes, then sliced cucumbers and tomatoes onto a plate and placed it, along with a pile of the standard-issue Georgian flatbread called *puri*, in front of me. Tariel was his name, and he began pouring shots of a distillate made from the grape pulp left over from winemaking. *Chacha* is similar to grappa, similarly wretched, and as commonplace as wine in daily life here. He toasted me in solemn, dramatic Georgian fashion: "This drink is for those who are working overseas. May they be successful and come back home with lots of money." Tariel somehow conveyed this via pantomime, inflection, and a few elementary words of English. I began the process that had become standard procedure for me, of getting drunk with shepherds in order to ingratiate myself with them and establish a rapport. I also like to drink, so this was not exactly a chore or obligation. We didn't share a spoken language, so I Google translated my words into Russian, which Tariel and the other shepherds took turns slowly reading off my phone. I felt bad shoving my phone in their faces when I should have been attempting to understand them. So I turned it off. Alcohol would be our shared language. After six drinks, everyone speaks the same language.

The three other shepherds in the shack didn't drink, and like most people who don't drink, they clearly knew how annoying drunk people are. They all had long beards, and when they said they were from Pankisi Gorge, it clicked that they were Kist Muslims who have cultural ties to the nearby Russian republic of Chechnya and are employed as seasonal shepherds in Tusheti. I liked these guys—they were quiet and serious, did the work without fuss. Tariel and I, on the other hand, got rowdy, spilling our drinks while eating young guda from this year that was pure white, and salted so heavily it stung the mouth. They helped me set up my poncho tent in the yard, and we sat in the fading purple light smoking cigarettes as the sheep came down a steep hillside, a white swarm pursued by an invisible shepherd who was yelling and whistling to urge them on. The whole gang of maybe three hundred plus some large leader goats crossed the creek and were locked in a night pen with a small section of the creek fenced into it, so they had access to drinking water. I laughed until my belly ached.

The place was breathtaking, and I was ecstatic, having found and been welcomed into exactly the scene I was traveling the world to find. Tariel and I helped gather some errant sheep, then stumbled back to camp with our arms on each other's shoulders, singing to the mountains as they were kissed by the last light of dusk. I fell asleep after tossing and turning a bit, waking to dogs barking like angry bears, the shepherds popping out of their raised beds at the corners of the camp, shining flashlights around the hills, hopefully scaring off the wolves that are always hungry for meat and blood. The Milky Way spun and churned into butter above.

The next morning, I was crusty, dusty, and a little rusty. Tariel was known to everyone in camp as Papa, and he was definitely the elder figure. He has a stubbly white beard and soft green eyes, and he reminded me of my father. For breakfast, he was taking shots of the foul gut rot again. I had to decline multiple times, shaking my hands and head adamantly. "Papa, no chacha!" It became an ongoing joke, a part of his name. Papa Nochacha. The sheep were stirring in their night pen, where they had slept huddled close together on a mat of compressed poop, wool, and straw that was nearly three feet thick and breaking off into the creek, like a melting iceberg of shit. A shit-berg. They were corralled into a series of long, rectangular pens made from rough-hewn birch poles, secured in the ground by stakes sharpened with an ax. The six rectangular sections were each as wide as a human's outstretched arms, and the sheep were pushed into these. A shepherd began at the end of a section, sitting on a stool, grabbing each sheep and quickly milking it over a wooden bucket placed on the muddy ground. The milked-out sheep would be pushed into the space behind the shepherd as he slowly moved forward, keeping the unmilked sheep ahead of him, packed tight.

I watched with a fair amount of shock as a sheep pooped directly in the milk and the shepherd nonchalantly scooped it out, not breaking his stream of banter. I'm pretty sure some pee got in there too. No big deal, it seemed. My eyebrows rose and jaw dropped as the shepherds dipped their lanolin-and-dirt encrusted hands into the milk, to lubricate them for milking. It was a powerfully captivating visual. Greasy, black-brown sausage fingers on swollen hands dripping with the white purity of innocence. The sheep had thick wool with some poop stuck in it, which fell in the milk along with the biological pixie dust of the sheep camp. The warm milk was then meticulously filtered. A large metal funnel was draped with cloths of various densities, then a pile of fresh-cut nettles was placed on top. The milk was poured over the nettles, and the hairs on the leaves caught a remarkable amount of debris, the finer stuff being captured by the cloths below.

The use of nettle to filter milk has a long history in many places, and the plant conveniently—and globally, it seems—grows close to barns and other areas of heavy impact on dairy farms, loving the overload of fertility and frequent disturbance. It accumulates the abundance of nutrients from manure, has medicinal properties, and provides a fiber that has been used to make cheese cloths. Nettle can also be used as a plant coagulant, and there is some evidence that even the brief contact of filtering can have an impact on milk microbiology, as nettle contains various antibacterial compounds. Again, we see the knowledge of healing plants and the use of abundant local materials interwoven with pastoral cheesemaking. I was impressed by the complete lack of concern about debris getting into the milk, and then the very careful filtration to remove it. Tariel proudly showed me what they had caught in their multi-tiered, reusable, and biodegradable filter system. The milk-soaked nettle, coated in wool and debris, was thrown onto a pile to compost and return its nutrients to the hills from which they came.

The milk now sat in a blue plastic barrel that was insulated with felted wool and wrapped in black plastic and duct tape. We poured enough nearly boiling water in to heat the milk back to udder temperature. As the other shepherds started moving the sheep out of camp, Tariel showed me the rennet, which they call *shabooshi*. A pail full of post-ricotta whey sat on a chair, a red chile and a literal sock full of chopped-up calf abomasa steeping inside. Tusheti was the first place I visited where the makers I met all made their rennet from locally processed abomasa. They used calf, I assume because they were widely available and the male lambs of these flocks were all raised to adulthood for their meat, the biggest income generator. I had seen the calf abomasa down in Telavi, the closest major lowland town. In the market, the butchers sold them whole, hanging on strings, salted and partially dehydrated with the "cheese" still inside. The shepherds would buy enough for the season and hang them in the rafters or attics of their shelters, where they were bathed in the smoke of the cookfire. They would make batches of rennet as needed, extracting the coagulating enzymes from the chopped abomasa in sour whey. Tariel explained how it impacted the flavor of guda by flexing his arms like a bodybuilder.

The rennet pail was stirred, then a mug filled to an eyeballed level. The rennet was chucked into the milk and stirred with a cheese harp made of fishing line stretched tight between short cross sections on a longer wood handle. I set my timer, and we checked the curd at sixty minutes, deciding we needed to wait. At ninety minutes, it was firm, so Tariel cut vigorously with the harp, breaking the coagulum down into curds ranging from corn-kernel- to walnut-sized. This was stirred for a few minutes, then the whey ladled off the top into a cookpot made from half of a metal drum. The whey was cooked into a ricotta called *kalti*, a staple camp food eaten fresh with sugar and stale bread brought back to life in the warm, moist cheese and whey. Very similar to the zabbina I had in Sicily, just less dialed to perfection. A coarse-handed, rough-hewn mountain ricotta.

From the blue barrel, the curds were transferred into a grain bag made of woven plastic, which worked quite well for draining cheese. This bag was placed on a wooden draining table and, with the curd packed into one corner, rolled and lightly pressed by hand to remove whey. As soon as it had knitted together as a cohesive, lumpy wheel, it was dropped into the bottom of a thick, nonporous plastic bag that came up to waist height. A cup full of salt went on top of this cheese, then the next cheese was placed inside, followed by another cup of salt. The morning milk had made four wheels, and these were stacked up inside the bag, which was filling with the whey coming out of the cheese. The transparent bag had its top closed tightly, sealing the cheese in with its salty whey. It began to dawn on me that perhaps guda was actually an ancestral brine-aged, feta-style cheese. A hybrid of the cheese-in-a-sack family and the white cheese I had seen in Kelmend. What would feta have been aged in before coopered barrels became common? Pottery perhaps, but it is heavy and fragile, burdensome to haul around. Maybe animal skins were the original, portable brine tanks.

Then it got weird, if it wasn't already. The bag of cheese was placed, with a few others, close to the open fireplace. They pulled one of the older bags over and showed me that it was bulging with gas. That cannot be good, I thought, and yet I could tell they saw this as desirable. They gave me a thumbs-up and indicated that they would wait for today's cheese to look like this, in about three days. They opened the swollen bag, burping its gas while pulling out a wheel and cutting into it. There were large, medium, and tiny gas bubbles inside the cheese, which looked to me like a combination of things that are generally considered defects. Some of the aromas hinted to me of clostridium and coliform bacteria. It's hard to say for certain which types of microbes were responsible, and there are of course gas bubbles formed by bacteria that are generally considered desirable, such as propionic, leuconostoc, and diacetylactis. Rather than showing concern, it appeared they were actually looking at these "defects" as signs of success, and encouraging them by keeping the cheeses warm. This bag of cheese was deemed ready to move on to the storage room, a tent full of similar bags of cheese on wood racks, with little salt deposits on the wood and around the openings where some of the brine was pushing out. The cheeses would sit here until the end of the season, when they would be brought down by truck to the lowlands and sold. The truck is a modern replacement for a horse or donkey, which once carried these strapped to its sides.

This cheese was a testament to the power of salt. The milk certainly had a thriving microbiota, and all this salt was being used to severely curb what would otherwise have been a chain reaction of life blasting off in all directions, some of them potentially unhealthy. I suspect there were multiple fermentations happening, with many genera of bacteria getting their gears in motion, digesting sugars and other foods into gases and metabolites beyond lactic acid before being impeded by the heavy dose of salt. That this could be considered a normal or desirable path for this milk to take fascinated me. The process was kept from entering the realm of dangerous or ruptured cheese by the power of salt, one of the few crucial ingredients in cheesemaking. The fact that the cheese sat next to the fire, in a bulging plastic bag, led me to conclude we were dealing with a wilderness of mixed fermentations that I'd never seen considered a part of the toolbox.

Salt, Milk, and Creek Water

This method of cheesemaking displayed a remarkable amount of dairy sovereignty. The cultures and rennet made by corporations that have seeped into most of the world's cheeses were not used. The only ingredient that potentially came from outside the country was the salt. Salt is not just crucial for the cheese; it is also one of the few inputs the flock requires. Transhumant pastoralists have to haul in all the salt that the herds, herders, and cheeses will need. This salt can also be used as an incentive for the animals, who crave it when it is not freely available. Hiking in Tusheti, I observed mysterious flat rocks surrounded by bare, dusty soil pitted with hoofprints. The mystery was solved when I walked with one of the shepherds named Abu as he dropped piles of salt on such rocks while singing loudly. As he did this, another shepherd would release the flock, which ran full throttle toward the yodeling man with the salt bag. The sheep relished it, fighting to get at the rocks and eating mouthfuls of the large crystals, then drinking from the river to wash it down. Abu would walk slowly up the valley chosen for that day, the sheep following, eager for the life-sustaining minerals in the bag he carried. As in Mongolia, ingenious means of persuasion based on an understanding of biological realities were used to coordinate the movements of massive herds without fences or herding dogs. You don't need to try to control or force your will on these sensitive, social creatures. Better to learn their language, and how their collective mind works, then steer the herd from within.

It is important to note that the shepherds were not drinking the milk. They were practicing what I consider to be the safest way to consume milk: ferment it as soon as possible, encourage bacteria to consume the sugars, remove some moisture in the cheesemaking process, and finish with salt. As in so many ferments, successful preservation is a matter of acidity, moisture content, and salt level. Safe food has little to do with shiny stainless steel, hairnets, and air filters. Those are modern contrivances, and the foods resulting from that approach are likely some of the riskiest humans have created. Ground beef and spinach processed in compliance with modern food safety laws have killed scores of people. It's not the cheese a neighbor or some shepherd in a shack is making that is getting us sick. It's the cantaloupe, chicken in plastic wrap, and jugs of dead milk at the grocery store you should worry about.

Another interesting microbiological point is that the shepherds and yours truly were drinking untreated stream water, even though the valley was crawling with cows and sheep whose night pens were right on the waterways. I can explain several reasons why this didn't get us sick. The animals were only here for three to four months a year, and then the valley would sit under snow. The herds were fit and healthy, consisting of disease-resistant heritage breeds that were culled heavily and undertook long, seasonal walks on rugged terrain. The water was flowing hard from springs and snowfields located nearby. Just because there are livestock in a watershed does not mean the water will make you sick. Not all poop is created equal; it doesn't necessarily contain harmful substances. In fact, poop is an incredibly valuable, integral part of how nutrients cycle. It is one of the greatest gifts our ruminant allies offer, and we have managed to turn it into a dangerous, toxic problem with our irrational modern livestock rearing systems. In the real world of microbes, soil, plants, and compost, poop is worth more than gold. We fear and despise poop because we fear and despise life.

Finding the Grail

It was amazing to witness all this microbial intimacy and layers of coexistence, but what I was actually looking for was cheese inside sheepskins, my grail cheese. A few people had mentioned the name David Shortishvili as someone who still aged guda in hides. Tariel told me he was in this valley, up near the top in a camp with many fierce dogs who would fight with every other pack and attack unfamiliar humans. He walked me up to the camp one day. The dogs mobbed us but did not attack, and we were invited in for a quick lunch with the shepherds. Grilled trout from the creek, old guda, cucumbers, tomatoes, stale puri. Lots of chacha and cigarettes. Then they took me to the attic of their wood shack, and there they were: arranged like a macabre wardrobe, dozens of sheep hides turned inside out, strung up on hangers like vests the sheep could wear on cold nights. The hides had a crinkly leather texture with two holes where the front legs had been, a head hole, and a large opening on the bottom, the sheep's waistline. The hides are removed in one tubular piece, like pulling off a sweater. They had been cleaned, salted, and lightly smoked as they sat up in the attic, a cook fire burning constantly on the ground floor below, smoke rising up through the floorboards.

They had a few hides soaking in water to be used that day. Once rehydrated, they resembled fresh hides again. They tied off three of the four openings, leaving one leg hole open, which they blew into. The hide stretched like an inflated ballon, and it was held up and inspected to confirm there were no holes. The open inflation hole was tied off and the hide turned inside out so the knots snugged up inside the leg and head holes. The sheep had been scissor sheared before slaughter, so it had short wool that was now on the inside of the sack and would be in contact with the cheese. The cheesemaking process was identical to what I had already seen, but they made the cheese in a wood barrel rather than a plastic one. After the cheeses had drained briefly, they went into the sheep hide exactly how they had gone into the plastic bag. A handful of salt, a wheel of cheese, more salt.

Most of the skin-sac cheeses I have heard of are made with the hair side out, from sheep or goat hides. In this situation, moisture can escape via the skin pores, which are a one-way street. For guda, the situation is reversed, so that the salty whey stays in, being unable to exit via the inverted pores.

In praising this cheese, I am not recommending that concern about milk handling and farm hygiene be thrown out. I'm not here to glamorize and glorify filth. It's probably a good idea to toss out milk that gets pooped or stepped in. Keeping the hair around the udder cut short and dusting it off with a rag before milking is helpful. When it's muddy and shitty out, extra precautions should be taken. The point is to use guda as an exaggerated example, and then work toward the center from there. In between guda and the ultra-sanitary, pasteurized, culture-with-DVI-starters, chemical-warfare approach is a middle path of microbial alliance and biodiversity. Deep terroir in cheese lies somewhere between the recklessness of addressing poop in milk with salt overload, and the opposite extreme of attempting to attain sterility while allowing the use of milk tainted with cleaning products and toxic products to keep your mozzarella white.

I ate plenty of guda and felt the opposite of sick. It wouldn't say it tasted good. I'd say it tasted . . . real. It tasted like the history of the people of this place. A convoluted fairy tale, full of passionate love affairs, longing, drunken feasts, and brutal feuds. It tasted like survival. It tasted like Tusheti, like sheep walking single-file along terraces carved by generations of their ancestors, following the map that is in their bones, back down the mountains. This cheese transcended terroir. It tasted like time, like time travel.

— CHAPTER TEN —

The Werewolf Cheese

SVANETI, GEORGIA

Maybe the wolf is in love with the moon, and each month it cries for a love it will never touch.

—ANONYMOUS

I rolled into Tbilisi from Tusheti and immediately got in another van heading to the west of the country. I was crammed between two large, hairy men, the type who take up all available leg space, and we sat there sweating on one another as the van, with well-worn suspension, jerked down the pockmarked highway. We crossed the Likhi Range that runs north to south, separating Georgia's dry east side from its humid west. The transition was dramatic; as we descended, everything became lush and green. The hot, sticky air was alive with the steady drone of unseen insects. Houses were built up on stilts in a river valley that obviously flooded frequently. Buffalo wallowed in marshes and mud holes in the red earth. I thought I heard a banjo ringing out from the hollow. We descended further into the subtropical thickness of the air coming off the Black Sea like a blanket of comforting suffocation.

I was heading for Svaneti, the most popular region of Georgia for trekking and other mountain recreation. It has the tallest peaks, a well-maintained network of trails, and tourist infrastructure. After spending the night at cheap hostel in a major crossroad city, I walked to the market and ate nectarines, plums, guavas, and buffalo yogurt for breakfast. I found the right bus, and we hauled off into the mountains. I had made arrangements to meet a family in Svaneti. I had been texting with Tamuna, the daughter-in-law of Ira and Zurabi Ansiani, who had agreed to host me for a few days. This family is possibly the last producer of Narchvi: one of Georgia's obscure and ancient cheeses that nearly went extinct during the

Soviet era. I got off the bus at the small roadside store that amounts to the town of Lahkamula. Zurabi was there waiting for me, a short man with a jovial face and thick eyebrows, who moved in frenetic bursts. Radiating vitality and good humor, he shook my hand firmly and bought me a Coke. Then we got in his little Russian-built, four-wheel-drive Lada hatchback and pulled out. As we did, the woman who ran the shop stood on the road behind us, making an intricate motion with her hands. I later realized this was the Georgian Orthodox sign of the cross, but it somehow looked more like a pentagram.

Reaching Zurabi's home, I was greeted by Tamuna, who showed me one of the vineyards, where the leaves were beginning to change color as autumn was already spreading in late August at this high altitude. Zurabi went off to do some chores with the men of his extended family who shared the large house. Svaneti has a distinct cultural identity, its own language (Svan) and cuisine. A staple food is fried cakes made of cheese and cornmeal. Cheesy corn fritters. At this house, they were golden crispy on the outside, still mushy in the middle, shaped like footballs. Highly comforting. They are made with *sulguni*, one of the four cheeses that were produced during the Soviet era, when the diversity of varieties made in the country was decimated, as dairying became a centralized, state-run industry. It is a salty stretched-curd cheese, like a drier mozzarella. After this snack, the chatter of a radio lulled me to sleep on the couch until Zurabi and the other men returned.

We sat down to a real lunch, just the men, as the women retired to the kitchen after covering the table with an inordinate number of plates piled with food. Zurabi launched into long-winded toasts in a serious, melodramatic tone. He was a *tamada*, a toastmaster—a bardlike role for respected male elders who are storytellers and keepers of history, and who more or less oversee or orchestrate a long meal with lots of drinking. They set the pace and regulate the rate at which everyone gets inebriated, since people seem to drink only when doing a toast. We ate fried onions and peppers with tender veal from the spring calves who were culled, their abomasa preserved to make rennet. I asked Tamuna about this, and she showed me the salted, dried organs whose enzymes would be extracted and used to turn the milk that the calves would have drunk into cheese for humans. The price of cheese is death. The act of eating is an act of killing. Even salads are filled with living, sentient beings such as carrots, lettuces, insects, and microbes, which are killed so we can stay alive. There is no such thing as a deathless diet.

The toasts were announced subtly. The air shifted as Zurabi straightened in his chair, his energy morphing from casual to ceremonial. I followed the lead of the men at the table and became silent, attentive. He filled all our glasses, then dove into standard tropes of youthful love, an end to war, blessings on one's family. (Later, he would freestyle off-the-cuff toasts suited to the moment.) He made a statement and then everyone clinked glasses. He spoke further, elaborating on his theme with theatrical hand motions. Then we clinked glasses again, nodding profusely. I began to raise the glass to my lips but realized he was still going. He sometimes went on for many minutes, everyone listening with glasses held at the ready. Finally, we drank. Whole glasses were drained, pitchers of wine chugged like beer. One wine was purple, another magenta, a third yellow-orange. I detected the distinct horse-blanket aroma of Brettanomyces, a common yeast that grows in beer and wine that is both celebrated and reviled. Some of the wines had sour-beer flavors; another tasted like a hard lemonade aged in a warm shed for a year. They showed me the pantry where they stored the wine, in carboys with huge pellicles of kahm yeast and God knows what on top, looking like the brine tanks of Kelmend. Forested islands of furry mold and cryptozoological SCOBYs left to do their thing. We siphoned from the sea of wine below the heathenish islands, refilling our pitchers and heading back to the table where Zurabi wrote poems in his head, inspired by divine grapes growing in his backyard.

Into the Misty Mountains

We got in Zurabi's car and he prayed while rubbing the beads and crosses that hung from his mirror. I supposed he was asking for protection before he proceeded to drive like a maniac, drunk, down narrow mountain roads, hunched over the wheel. This was the greatest danger I'd been in while traveling, riding in a car with a drunk driver who felt he knew the road so well that he could and should drive at top speed. It was nerve racking. I could have refused, but I somehow felt committed to going wherever he intended to take me next, and so I said my own prayers and tried not to watch.

It was almost dark as we turned off the highway and began driving up logging roads into the heavily forested mountains, with mossy pine trees and a lush undergrowth of ferns, brambles, and bushes. Twin Peaks vibes, like where I grew up in western Washington. Wet . . . dark . . . mysterious . . . fungal. Bigfoot and some elves could have happily coexisted here. We kept winding up and up, for an hour and a half, all bumps, tight corners, carsick-type driving with no sign of human habitation in the spooky hills. At about ten p.m., consoling lights appeared out of the darkness and we crawled into a small collection of homes and farm buildings in a clearing. About a dozen people came out to greet us: Ira and Zurabi's children, grandchildren, and in-laws. An extended family whose exact relations I never could figure out. They shook my hand, and Zurabi introduced me as "America Chicago." I'm not sure why, but you don't get to pick your nickname, so there I was. We all filed into a cinder-block house with exposed lightbulbs and loud, garishly colored plastic wall coverings. Ira was cooking up a storm and stoking a fire to keep the cold mountain night at bay. She never ceased working. A language barrier existed, but it wasn't high. Sometimes people who share a language have larger barriers between them. Communication is as much about intent and effort as it is about words. When two hearts want to connect, they find a way.

We proceeded to drink chacha and eat a midnight meal of grilled trout and cheesy, creamy potatoes, along with the standard cucumber-tomato salad served with Svaneti salt, an appetizing blend that contains cumin, coriander, chili, and fenugreek. I was surprised to find so many people up here, and they were intrigued by my presence, staring at me and whispering to one another while suppressing giggles. I stared and laughed back, enjoying the attention. It was good to see children and teenagers involved in the work of dairying and off-grid living, participating in the seasonal transhumance in which cows, dogs, horses, and hogs were brought up to this tree-line mountain camp. They spent the summer isolated up here, working with animals and one another in the middle of unpeopled land, far from paved roads. I slept in the main bedroom, which was full of beds, yet no one else slept there. Ira and Zurabi slept out in the living room, which seemed strange but is par for the course of Georgian kindheartedness. There is a saying here: "The guest is God." So they give you their bedroom, feed you, and would never dream of letting you help cook, clean, or participate in farm chores. Guests don't have to earn their keep, pitch in, make themselves useful. It would actually be an insult to try.

Milking happened before sunrise. I heard Ira go out and start waking up her cows like a mother waking children who were late for school. I lay there a minute, fearing the cold, then limped outside as the stars slowly yielded the stage to the newborn day. Ira was careful, wiping the cows' teats and udder thoroughly as she hand-milked the herd of about fifteen, leaning her head into their ribs for support and warmth. The majestic horned cows had a diversity of coat patterns and colors. They belonged to a loosely defined breed called Georgian Mountain cattle, one of the smallest of the dwarf cattle breeds. Their size lends agility and the ability to maneuver on steep, rocky terrain where a heavier, taller cow could hurt itself. I've seen small, nimble cows in many mountainous regions; nimbleness is a trait that's been selected for because it works. The large modern breeds couldn't cut it in the real world; they're designed for a pampered life in artificial environments. Like other heritage livestock breeds, Georgian Mountain cattle display a high resistance to disease and parasites, and the ability to produce milk and put on weight in harsh climates with poor forage. I have heard these same traits listed for so many breeds. It's apparent that these are not special characteristics; they are innate animal traits. These strengths have been lost in our industrial systems by breeding narrowly for high production, large size, and fast growth. It was impressive to watch these athletic cows plummet gracefully down rough trails, maneuvering along steep hillsides and crossing rocky creeks with ease. The most rowdy, goatish cows I've seen. Built to move, thrive, stay healthy, and provide their gifts without requiring much monetary investment. That is true efficiency.

Full Moon Rising

Ira took the morning milk inside and mixed it with the evening milk that had been kept in a can sitting in a bathtub with cold water flowing through it constantly. A creek had been diverted to run right past the house, then gravity fed inside. She heated the milk in a large pot and added her homemade rennet, which resembled what I had seen in Tusheti: a bucket full of sour whey, dried abomasa from her herd cut up and placed in a cloth sack, and a few chiles floating on top, probably playing a preservative role or deterring insects. She partially filled a cup and put it in the milk that was heating on top of a woodstove, where she also baked bread. She moved the pot off the heat and left it to sit until the milk coagulated, at about one hour. Ira broke up the curd with one hand in a swirling motion, her fingers curled into a whisk. Then she periodically stirred it with a wood spoon as it sat on the hot stove, slowly heating up as the curds became firmer, releasing their whey. She shaped the cheese as two wheels drained inside colanders, flipping them often to give a uniform shape and ensure proper drainage. Just milk and rennet from her cows turned quickly into cheese—this was one of many farm chores, the final step of milking.

Ira showed me the cheeses she had made in the past week, which sat unsalted in a tub. These wheels had various flavors, textures, aromas, and levels of acidity. The three-day-old wheels smelled pleasantly yeasty, with notes of banana and bread. On day four, they had stronger, less appealing yeast aromas starting to go overboard, with something bolder and meaty starting to growl. The cheese was a bit slimy, the color going from shiny white to a dull, non-reflective off-white as a fungal mat began to bloom. The five-day-old cheese smelled and looked like what even I would have to call a bit off. And on day six, it looked like it might crawl up my leg and sink its fangs into my neck. It smelled bad, like meat that was starting to turn. I was scared of it, and thought, *Why would you ever let a cheese just start rotting away like that?*

These cheeses were still unsalted. Without salt, cheese can start going down paths that resemble decay rather than preservation. Inside a warm kitchen, cheeses rapidly transform as yeasts begin growing on the surface, imparting aromatic compounds as they digest the lactic acid on the surface of the cheese. Rapid enzymatic reactions occur as the enzymes in raw milk, in rennet, or released by microbes alter and break down the cheese, and microbes are allowed to run amok, without salt's selective sedation. This expedited pre-ripening of the curd creates an enzymatic booster shot that sets the trajectory and accelerates the ripening process as the final cheese ages, with salt finally reining in the chaos. Many cheeses start fresh and innocent, then slowly unhinge and get quirky, fierce, or interesting as they age. Narchvi comes rushing out of the gate as a nutcase firing at random, then matures into a wise old storyteller with a crazy past and beautiful tales of love, loss, and finding God.

Ira took a two-day-old cheese and cut it up, then stretched the pieces in hot water into sulguni. Those wheels had fermented into the stretch window, and she made only what was needed for the dining table. That was just a sideshow, a practical way to get a fresh melting cheese for immediate

The ger in the valley of the yaks where I spent eleven days. Old Bill the horse, a satellite dish, and a solar panel complete this scene typical of the Mongolian countryside.

Morning milking of the yaks. The calves are tethered to a staked long line. The mothers are hobbled and babies released to suckle before a human steps in to milk.

A coopered wooden milking pale with tethered moms in background.

The boy stands between Burmaa and the mother's head so she can't see who is taking her milk. This is a cow-yak hybrid.

Heating milk and forming the curd for byaslag, a pressed cheese that is dehydrated.

I sit next to Burmaa and her family on arrival in the valley of the yaks.

Aaruul, the dehydrated yogurt cheese, is placed on boards on the roof of the ger, where it dries in the sun and wind.

The herd of Saanen goats at the lower farm, with a panoramic view of the slopes around Gran Paradiso, near the border with France.

The evening's cow milk is left in cold, flowing spring water outside the upper alpage.

The cheesemaker at the upper alpage of Azienda Agricola Arpisson in Valle D'Aosta, Italy.

A wheel of caprino made with kefir starter grows a natural rind in the cantina.

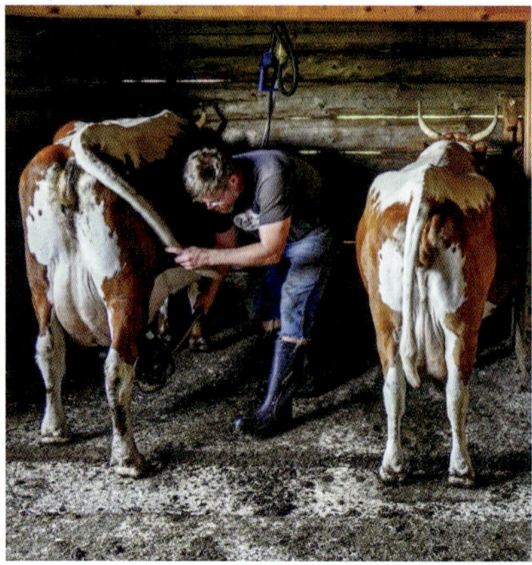

Roy getting a cow ready for milking in the old milk shed at Braskerudsetra.

Two Telemark cows, a heritage breed from Norway, at Braskerudsetra. One has gold tips on its horns.

Roy empties firm curds for pultost from a copper pot.

Butter is also pressed in the trough carved from a single log.

A hinged wooden block form shapes brunost into a pillar embossed with Nordic patterns as it cools.

The milking barn and creamery at a seter called Grindal Ysteri with a green roof made of sod blocks.

Seter rømme, a sour cream that is served with pancakes or pastries, berries, and sugar and washed down with black coffee.

The seventeen-year-old one-horned matriarch of the herd that summers at Krstenica above Bohinj Valley, Slovenia.

The milking shed with floors and walls that can never be sanitized.

Small rounds of Krstenica sir sit on aging shelves in a room connected to the creamery.

Alfonso propped on his trusty ax as we watch over the herd on the rocky ridge above Jerte Valley in Extremadura, Spain.

A Verata goat displays the chestnut coat with black ridgeline and legs that is one color pattern within this genetically diverse heritage breed.

Alfonso cuts the coagulum of the cheese called *queso* by stirring rapidly with his hand.

The lower milking shed of the Cangemi family in Sicily.

The sheep head out for a day on pasture.

The handwoven reed baskets that give Pecorino Siciliano its final shape.

Martina pouring hot scotta onto the curd that will be stretched into vastedda, a rare stretched-curd sheep-milk cheese.

Ewes and their lambs graze on pastures recently freed from snow in the village of Lëpushë below the Accursed Mountains in northern Albania.

The table where I first tasted Mishavinë, on bottom plate with fork. Above it is mish i thäte, a dried smoked ham, and to the right is white cheese, which is similar to feta.

Fran holds up the wooden vessel that will be filled with milled curd and coated with butter to age into Mishavinë.

A young cow with an evil-eye amulet that contains mustard seeds and garlic cloves to protect the cow from negative influences.

A colorful herd of goats crosses a bridge to reach the open grazing grounds on mountain slopes.

Dried and smoked calf abomasa hang in the attic of a shepherd hut in Tusheti, Georgia.

Tubular sheep hides that will be stuffed with cheese wheels and salt to create the vessel in which guda ages in a brine.

Tariel and the other shepherds milking in the partitioned corral.

A shepherd watches the flock of Tushetian fat-tailed sheep as they descend a hill.

The heavily grazed area and maples around the summer farm in Svaneti, where the herd of colorful Georgian Mountain cows are brought each summer.

A wheel of guda is placed inside a sheep hide with copious amounts of salt.

Ira Ansiani salts and mashes the week's worth of cheese that was put through a meat grinder and now sits in a stock pot.

After draining, the milled curd is packed in a wooden box and aged with a lid on top.

Finished Narchvi, with layers of creamy, crumbly textures and a pink hue on the outer edges where it was in contact with the porous wood.

Horned, multicolored, petite Georgian Mountain cows.

Iia stretches curd into a single long rope in a pot of hot water.

Strings of Tenili dry on a rack.

Remains of a dairy-centered meal. Clockwise from top right: matsoni, a Georgian mesophilic yogurt; butter; a cheese aged in sausage casing; Tenili aged in cream; fresh hunks of white cheese; and a bowl of sour cream.

Canarian sheep with lambs graze on the green flanks of volcanic Gran Canaria, in the Canary Islands.

Cabra Palmera goats on La Palma.

Smoked Queso Palmero in various formats sits on shelves at Los Barbusanos.

Wheels of Queso Palmero that were salted with local sea salt on the same day they were made.

The marshy grazing grounds near The Farm at Chennai, India, with white cranes and the encroaching city behind.

The buffalo and some cows wallow in a salty pond.

The buffalo herd being brought home through palm trees after grazing.

Wheels of the raw-milk, smear-rinded buffalo tomme.

Brown Swiss cows graze the pastures above the village where I stayed with Anton in Bregenzerwald, Austria.

A pyramid of wooden gebsens in which the evening milk sits overnight, allowing cream to rise.

Anton scoops up a wheel's worth of curd while holding the cheese cloth between his teeth.

The wood barrels containing sour scotta that is used as the acid to make ricotta and as the sole cleaning product in the creamery.

consumption out of a larger batch. The main act was a cheese with the enchanting name Narchvi. It is a milled-curd cheese pressed and aged in wooden boxes, similar to Mishavinë in many ways. Ira put the weeks' worth of unsalted wheels that ranged from two to seven days old through a meat grinder, the resulting gooey crumbles mixed together in a large stockpot. This milled curd sat in the pot overnight. I was delighted and alarmed to see it had started to grow a rind by the next day. I've never seen a cheese so primed to grow fur and turn into an animal. The surface was exposed to air and had a visible yeast layer starting to grow with a vaguely pinkish hue. The stuff under this did not smell pleasant, and was releasing some heat. The aroma reminded me of durian and overripe papaya, along with a whisper of meat rotting on a beach. Funky in a bad way, what I would call a failed experiment and throw away. It's a vague, ill-defined, subjective line that exists between funky and rotten. Between delicious and disgusting. Civilized and wild. It was in this murky purgatory, growing moss in the moonlit forest, that the cheese I was looking for prowled. I wanted to catch that werewolf cheese, at the moment of metamorphosis. A lactic lycanthrope. It might eat me, but I needed to see its tortured eyes, just to know it was real.

Finally, as I sighed in relief, the ground-up composting cheese had salt sprinkled over it in a large bowl. Anointing the beast cheese with saline salvation. It was mixed by Ira's seasoned hands with a squeezing motion that sent it out of her fist as a paste that felt like a dense, wet shortbread dough. Ira was a strong woman, and she mashed the cheese with gusto while wearing a cool bandana on her head, babushka-style. She kind of ignored me, letting me take my photos, offering me tastes of everything, not letting words get in the whey. Allowing her cheese to speak for itself. I liked how she calmly did her job in a kind of focused trance that many talented cheesemakers seem to enter as they commune with their medium. Once the cheese paste was salted to taste, it was packed into the same grain bags used in Tusheti, then stood upright in a large bowl to drain overnight, a greasy, yellow whey seeping out. I tasted this whey and the salted curd. Both tasted quite nice with a meaty umami depth, not like the off-putting aromas that had greeted me before. It was a combination of familiar flavors: butter, sour cream, fruity yeasts, umami-rich cured meats; and new ones I couldn't explain, since they were so novel. The only word that came to mind was *crittery*. The next day, after morning milking and cheesemaking, Ira moved the curd from the grain bag into its final resting place: a rectangular wood box. She packed it in firmly and fitted a square lid snugly on top. This box was placed in a cold storeroom with gravel floors; pressure was applied to the lid by a coil spring from a car's suspension and a tire jack. This was hands down the most redneck, and resourceful, cheese press I'd ever seen—and remains so to this day.

After aging a few months to a year, the cheese was ready. Ira had a two-month-old box that had been partially consumed that she kept covered with a cloth to keep it from drying out. She removed a large cube and placed it on a plate. Like Mishavinë, it had similar red and white discolorations on the edges where there was contact with the porous wood, the slight air exchange allowing yeast and bacteria to form a quasi-rind layer with little divots full of *Geotrichum* and some blue mold intruding through the borderlands. The cheese crumbled apart easily into shards of various sizes, textures, and colors. I mashed it in my fingers until it became buttery. As I ate it, my mouth watered. The textural variety was highly stimulating, the salt level perfect. The textures and flavors shared characteristics with Mishavinë. This was a bit more extreme, though—deeper, wilder, and meatier. It was highly savory, almost too rich on its own, like concentrated mushroom broth or liver paste. An umami bomb. Half cheese, half wolf.

Narchvi reminds me of a famous British cheese called Kirkham's Lancashire, which is made by mixing two- and three-day-old curds that get some yeast growth before being milled and salted. This curd is pressed into large truckles (a cylindrical shape that is taller than it is wide), which ripen quickly, likely due to that two or three days of sitting unsalted, letting a breakdown kick in that accelerates the process of flavor development. Kirkham's has a delectable texture described as a buttery crumble. Narchvi has a similar texture and juicy mouthfeel, but everything about it is less refined, less polished. Less British. I developed a story to explain the connection: Kirkham's has an eccentric cousin who never fit in and left the British Isles to start a family in the mountains of Svaneti, embraced the local culture, drank a bunch of chacha, and became a little rough around the edges. It lost its proper manners and became bestial, as wild as the mountain streams, growing a long beard. Narchvi is like Kirkham's gone feral and a little crazy out in the boonies, living in a little wooden box and running barefoot through the forest with unkempt hair, talking to the trees. Narchvi pushes the logic and beauty of a slow fermentation and the mixing of multiday curds to an extreme, to the boundary between fermentation and putrefaction. It goes right up to the edge, then pulls back, and somehow deliciousness results. Tantalizing, carnal, with a vitality that reawakens a palate weakened by grocery store mundanity. Like Kirkham's, the flavor of Narchvi is mainly buttery sour-milk love in your mouth, but there are also subtle, dark overtones of bitterness, liver, overripe fruit, prosciutto rind, and death.

So much about this cheese is anomalous, rule bending, boundary dissolving. Cheese is typically made on one day and salted the next. In the modern industrial approach, you take milk that was maybe stored cold for two or three days and make one large batch of cheese. This cheese is forced to ferment quickly, as a safety mechanism. A rapid fermentation is almost always a safer one, and is also a compensation for microbially suspect or impoverished milk. But rushing the process doesn't often lead to striking, singular flavors. Safe cheese is not necessarily delicious. Many of these cheeses I was homing in on, the ones that really seemed to have something to say, that had tangible terroir, were marked by a slower fermentation, with two or three days elapsing before the cheese got its blessing of salt. This can be compared to baking sourdough, where a slower, toned-down fermentation often leads to greater depth of flavor and nuance. This approach can lead to interesting expressions of milk and microbes that do not have the chance to manifest when the process is rushed. These more subtle, nuanced flavors are drowned out in modern industrial cheeses by a hearty dose of starter culture microbes, which bombard the milk with a few strains of LAB. Graham Kirkham, cheesemaker and head of the Kirkham's team, told me he adds a "just a whisper" of culture to the milk. With these intentionally restrained and delayed fermentations, there is more that can go wrong. And more that can go right.

Ancestral Milled Curd

A suspicion grew in my mind that Narchvi and Mishavinë could be considered ancestral milled-curd cheeses, and that they are linked to the skin-sac style discussed in the last chapter. The coopered barrels that Mishavinë are aged in and the wood boxes of Narchvi have a history going back only so far. What would these cheese have been aged in before this? In Kelmend, I was told that Mishavinë was formerly aged in a sheep hide. I suspect the same is true of Narchvi. The practice of making a simple cheese once or twice a day and storing them until you have enough to fill a hide is likely incredibly old. It allows you to make a large cheese that can be stored long-term without adequate cellar conditions of low temps and high humidity. To get a cheese to age long-term, you have to remove a certain amount of moisture from the curd. If cooking the curd is not possible, the moisture can be forced out by physically breaking up and mashing the curd. The salt is dispersed throughout by hand, then the whole mess packed into a semi-porous vessel that becomes the cheese-aging space, preventing excessive dehydration and rinds gone wild. Hides and pottery are two materials that long served as the vessel before wood and then plastic came along.

There are fascinating milled-curd cheeses with long histories in many parts of Europe, such as the Spanish Gamonéu, French Cantal, or the Italian multiday curd Castelmagno. The British Isles likely have the largest diversity of this style. I'm not saying they are direct descendants of the ancestral milled-curd cheeses aged in hides. Similar approaches have probably been stumbled across independently, or cheese technologies have spread inadvertently in a manner that is impossible to definitively trace. My approach is to celebrate the diversity. I see far too much fighting over who owns a particular style, too many attempts to draw conclusive histories of the spread of technologies, arguments that are often colored by nationalistic overtones. Narchvi is a cheese of Svaneti. Mishavinë of Kelmend. But the approach—of mixing multiple days' worth of cheese together and packing it into a skin, a box, or a lovely truckle wrapped in cloth—belongs to humanity. The Narchvi made by Ira is the cheese of her home, her people, the creeks that flow down the mountain, the pastures that the mountain cows traverse to find their feed. But a facsimile could be made anywhere. It might not work well, and it would likely take years of trial and error to get it on track. It wouldn't be Narchvi. But it could be something equally strange and delicious. With its own name, and maybe, after enough time alone up in the hills, its own terroir.

Wild Tending with Livestock

I hiked up a glacial-carved valley full of summer flowers to the base of jagged snow-clad mountains shrouded by gray clouds. For much of my life, I have been drawn to such places, wilderness areas free from human interference where I can soak in the beauty of nature, be immersed in the subdued symphony of birds, insects, and water flowing over stones. Here, far from cities and suburbs and even small towns, without any humans around, was the real world, where I felt at home. The problem is you can only stay out there for so long before you have to go back. You go up, then come down. My perspective was infected by the sense that the wild was other, something that I wasn't a part of, a state that only existed in the absence of humans.

Something had shifted inside of me over the course of this trip. Now wilderness felt like it was missing some crucial element. I looked back down the valley to the camp, a collection of small wooden houses with smoke rising from chimneys. I could see children running around playing, and I could hear fragments of their laughter carried by the wind. A herd of cows moved almost imperceptibly across the open meadows next to a steep creek as they grazed without fear. That was where I wanted to be. The camp looked, felt, sounded, and smelled like home. Not my home, but a home for people and animals. It was in between wilderness and town life. A human habitation with shelter, warmth, conviviality, and plenty of food that was not separate from the landscape surrounding it. Perhaps these pastoral lifeways and settlements appealed to me because they offered a third way, a means of building a life, home, and community within the natural world, as a part of it. Our lives can be as beneficial and perfectly natural as those of beavers, deer, or fungi. We, too, have a seat to fill in the great orchestra.

The land surrounding the camp was open and covered with plants commonly seen in areas many would call overgrazed. Curly dock and sheep sorrel were the main ones, which I assumed the cows were not eating, although they are edible for humans and quite tasty. Some flowers and herbs grew here (nettle, yarrow), as well as tall maple trees. It was a zone of heavy impact and frequent disturbance, where the cows were brought through every evening before milking. Walking out from here along the trails that the cows were herded down, I entered a second ring that was less impacted, but still free from pine trees and heavily terraced by cow paths. A mixture of plants grew that resembled the meadows I had walked through far from camp, but with a different composition of species. They actually seemed taller and healthier than the wild meadows, perhaps from the nutrients introduced by the cattle. Where the cows were generally grazing was a broader third ring, with more grass and desirable fodder. The forest of conifers and understory crept into and blended with the edges of the grazing areas in a fuzzy transition.

The human-livestock influence on the landscape radiated out in concentric rings, moving from the human habitations and the heavily impacted zone through gradients of less impact, until it blended back into the wider ecosystem. There were no sharp divisions. No borders. No fenced property lines: private on one side, protected state land on the other. The overgrazed zone was just a small dot on a large canvas, surrounded by rings of influence rippling outward. Set within an otherwise rather monotonous botanical painting of forests and wild meadows, this gradient of human impact increased diversity, adding colors and textures to the mosaic, increasing the edge zones. I noticed there were more birds here, more fertility. This corresponded with previous observations that the transitional zones between dense human populations and uninhabited areas often have more wildlife than farther into the wild areas. The wildland-urban interface is a zone rich in birds, insects, animals, and ecological diversity. The interface supports a greater concentration of life than either side of the line.

The edge is where the action is.

Maybe little bits of overgrazed land can provide unique habitats, where specific biotic communities can flourish, not found in the surrounding "natural" landscape. Maybe *overgrazed* is the wrong term—too simple, subjective, devoid of context. Heavily disturbed areas created by occasional landslides, fires, or the movement of large ruminants create potential, and if they are allowed to regenerate, life flourishes in these openings. Nature loves and needs disturbance, which creates space for seeds to sprout, for experimentation, for novelty. Maybe what was missing from the wilderness I had walked through in that valley far from camp was this light caress of the human hand, and of the hoof. Maybe the disturbance to the land from our living, our raising of livestock, our growing and gathering of food can be beneficial, creating more space for life. Maybe natural, functioning ecosystems can include and be served by human-livestock-microbial alliances. Maybe the earth needs us as much as we need her.

— CHAPTER ELEVEN —

Layers of Terroir

SOUTHERN GEORGIA

Borders are set up to define the places that are safe and unsafe, to distinguish us from them. A border is a dividing line, a narrow strip along a steep edge. A borderland is a vague and undetermined place created by the emotional residue of an unnatural boundary. It is in a constant state of transition. The prohibited and forbidden are its inhabitants.

—Gloria E. Anzaldúa, *Borderlands*

What drew me to Georgia and kept me there for three months of pinnacle cheese trekking was the range of anachronistic techniques found in the country. These techniques represent preindustrial approaches to making cheese that don't fit into the standard concepts of what it looks and tastes like from a Eurocentric perspective. To treat European cheese as the pinnacle and to hold all other expressions to that standard is clinging to a racist, imperial ideal, and reflects the ongoing inquisition. The barbarian cheeses I seek aren't created by a person who develops a recipe, mixes specific strains of microbes into milk, and maintains strict control over the process. They come from an impulse that is older and deeper than the cultural wound that masquerades as advancement. For thousands of years, humans have adapted their foods and built spaces in accordance with their surroundings and found ways to turn milk into cheese that fit their lives. The techniques may have come from somewhere else, been borrowed or passed down, but trying to classify things as native or introduced misses the point. Foodways move and traditions are alive. They flow and adapt. The cheeses I seek aren't produced and marketed. They aren't made to appeal to the masses or be approachable. They spring forth from the ground like trees reaching toward the sun.

In two villages in the south of Georgia, near the border with Turkey, a stretched-curd cheese called Tenili is made from skim milk. A single rope of cheese is slowly stretched in hot water, then pulled into very thin strings, which are hung to dry for weeks as yarnlike bundles. Finally, the cheese threads are coated in sour cream and packed in terra-cotta pots, to be consumed fresh. In the past, it was probably aged like this. This fits into a historical practice in much of the world of transporting and storing all sorts of foods, oils, beverages, and animal products in tall, amphora-style pottery vessels that could be buried in the ground to stay cool for longer storage. The practice has survived in Georgia with the prolific use of *qvevri* in the making of wine. Materials matter, and they can be a crucial component of a cheese—of its recipe and its roots. When the makers of a cheese insist on using the traditional, natural materials, it also supports the artisans who weave the baskets, make the hemp or nettle cheesecloths, shape and fire the pottery forms and storage vessels. The copper pots; the wooden vats, forms, or aging vessels; the animal organs or hides used in cheesemaking: These crafted materials, like the cheese, are born from a place and its history. They are an aspect of the cheese, and in replacing them with something cheaper that can't be made or repaired by skilled human hands, the power of the cheese, a piece of its terroir, disappears. The tree is severed from its roots.

I heard about a village where multiple families were still milking the old cows and making Tenili, in an organized effort to revitalize the cheese and their rural agrarian homeland. I rode buses out to meet Ruslan and Iia, who work together on this effort and drove me up to the village of Andriatsminda, where I would spend the next few days. The small town clings to a ridge with terraced fields spread for miles above it. These fields are no longer planted with crops but instead are used as hayfields and pastures, or else are slowly reverting to forest. There was a young girl in the village who spoke English and was excited to act as an interpreter. After I got settled in and ate, she showed me around the village with her friends. Many of the old stone houses—patchworks of materials, colors, overlapping time periods, and layers of plaster and paint—sat empty, like so many of these fortresses where dying cheeses make their last stand. The people who remain here are proud of their small subculture and the dairy foods that are the foundation of its cuisine. The children were enthusiastic guides who loved their home, and their presence in the village was its hope of survival. They showed me the tiny Orthodox church, which was locked up, but we all made the sign of the cross and continued. At the end of the ridge, there was a tall cross between the valley below and the village on the hill, with mountains and an international border behind it. Right on the edge of a nation-state, this little pocket of culture had escaped the industrialization and development that have been the death of food and political sovereignty on much of the planet. The edges, the mountains, the borderlands, the porous membranes. If countries are cheeses, then these borderlands are the rinds, where exchanges occur, where things get interesting. Where life flourishes, peaks in biodiversity, rapidly transforms, and finds a way through, under, or over the wall.

Ruslan is a jolly, hardworking farmer, with bright, youthful eyes that overflow with energy and love for his village and the agricultural work that roots his culture in the land. He told me his people and their cheese are indistinguishable. Without Tenili, they would lose their identity and cease to be. I followed him up the hill, past the last houses with their barking dogs and broken-down vehicles, up to a geriatric barn where cows were milked. Coming up from a creek bed, looking like Pleistocene cave art, the striking herd of curved-horn cows had colorings and patterns that seemed painted on from the palette of the soil and rocks of these hills. Many would call them Georgian Mountain cattle, but they look a bit different from those I saw in Svaneti. Modern animal breeding often gets dangerously close to inbreeding, and genetic uniformity is a part of creating standardized, commodified breeds. This parallels the narrow breeding of plants and loss of genetic variety in food crops. The genetics of this herd were obviously diverse, judging from the variations in color and coat pattern. Pastoral breeding is aimed at maintaining a type of animal while bringing in enough new genetics to keep the breed robust. The concept of distinct, static breeds falls apart in the real world, in the pockets of sanity where life is not yet seen as a thing to be dismembered and placed into distinct categories that can be labeled. These cows represent another worldview. A way of living where things are allowed to move. When you replace the cows with "improved" modern breeds and replace the wood and pottery with plastic, layers of the cheese evaporate. The story loses some of its meaning.

A herder had been sent to gather up the cows for milking. He didn't spend the whole day with them but checked on the herd a few times via binoculars from the prominent hill where the barn sat. He drove them from the bottom, as Ruslan whistled and cajoled them from the top. They glanced at me as they passed on the terraced trails that they had etched into the hillside. Petite and strong, elegant and dainty, they were seasoned climbers, rough and scrappy royalty. Mountain queens. As they filed into the barn, a bull stopped to stare at me, sizing me up, squaring off. Ruslan saw this and ran at the bull, waving his stick and yelling. Acting like a bull. He warned me to keep my eye on the old boy, who was protective of his harem of cows. Being challenged by a bull is kind of a universal, prelinguistic experience. It doesn't matter where you grew up. You might not know what to do, but anyone can read dominant male power displays and interpret showboating bro behavior, no matter the species. There is nothing wrong with intact male animals acting this way. Nothing could be more natural, and the popular shaming of this behavior and looking down on the unbridled sexuality of livestock is a denial of life. A culturally ingrained fear of sexuality.

A few dry cows and the bull were pushed into a pen while the milking cows headed inside to line up in their customary spots, assigned by the herd mind. The cows were secured in place on either side of the barn, with a poop trench behind them and a feeder in front, where they ate hay in the winter. It took hours for the herd of about twenty-five to be milked out by machine; they seemed to be holding back their milk and dropping it very slowly. Well after dark, we loaded milk cans into a truck. The cows lay down to sleep in the barn heated by their bodies, permeated with their scent and essence. I imagined cows had slept here for decades, the walls thriving with microbial rainforests that permeated the wood, inhabited the soil floor. That barn carried aspects of the multifaceted terroir puzzle that I was struggling to put together, and that I can still taste today.

Tenili

The bottom story of the house had been converted into a creamery, with a storeroom that contained a wood-fired bread oven. The evening milk sat in this cool storeroom overnight, then was skimmed of its cream with an electric centrifuge. Some cream was used to make butter, but most was reserved for the sour cream that is a defining part of Tenili. The morning milk was skimmed while still warm, combined with the evening milk, and put into a collection of plastic bins. The milk was set with an inexact amount of powdered FPC rennet, and once firm, the coagulum was broken by hand and stirred periodically, then removed into plastic colanders and left to drain overnight. Fast and casual, it was the same basic process that by now I had seen dozens of times before, which could lead to radically different cheeses, depending on what would be done from here. Depending on the guidance of human cultural practices, recipes, tools, and techniques.

The next morning, Iia cut into one of the wheels and smelled it, nodding approvingly. *It's ready*, she told me with her whole face. She was cool, wearing an all-black outfit and rain boots, barking orders and doing ten things at once, and somehow putting up with me, smiling when I was clearly getting in the way. She had that sense of urgency one needs to do this work, to direct the orchestra and perform the heavy lifting of daily cheesing. The fresh chunk in her hands smelled of sweet-and-sour milk, the aroma of a cheese that had fermented into the stretch window. The same basic aroma presents itself wherever and whenever a cow's milk cheese is sour enough to stretch, with a pH between 5.4 and 5. Iia had already fired up a powerful propane burner under a pot of water, which she dipped a piece of cheese into with a ladle. She held it between her sturdy, skillful hands, stretching it out like a piece of toffee. With her arms at full spread, it elongated and sagged with gravity's pull. She nodded and smiled. *I told you so*, she said with her eyes, then cut the wheels into strips.

Iia tossed a handful of the strips into the hot water and stirred slowly until they began melting, then she gathered them together as a homogenous blob. She picked this up with thick, working hands adapted to these temperatures and flattened it into a disk, breaking through the center so she had a four-inch-thick donut. This donut expanded and thinned as she ran it through her hands. It nearly touched the floor, so she doubled it up on one arm and started spinning her arms in fast circles, the band of cheese cycling through the hot-water pot to keep it stretchy. The band elongated rapidly, and she kept doubling it up, so that it began to cover her arms like a coiled rope, a single thin band of cheese that had to be over 160 feet long. She kept going, faster and faster, until the rope was shoelace thin. It was mesmerizing to watch, a blur of serpentine motion. She dipped the whole bundle in cold water a few times, to stop the stretch and prevent the strands from sticking together. She shook it and wrung it out, then dropped it in brine to absorb salt for an hour before hanging it to dry.

I had never seen cheese stretch like that, into such thin strands that stayed separate. It seems reasonable to assume this was in part due to the reduced-fat milk. The day after stretching, the salted strands were placed on a table and three elderly ladies from the neighborhood pulled them apart into even finer, weblike threads, some as thin as hair. They had obviously done this a lot, and sat there chatting and joking in exactly the same way old ladies do in any congenial neighborhood. It's like pulling apart string cheese but really obsessing over getting the strings as thin as possible. These strings were very long, and the women didn't fully separate them as they created what looked like bundles of yarn. I had no idea you could do that to milk. The bundles were draped over bars on a stainless rack and left to dry in the creamery for up to twenty days. During this time, they got yeasty and likely underwent some of the enzymatic processes outlined for Narchvi, but in a more restrained manner, as the cheese was salted and in a totally different form. No mold appeared to grow. The hot-water stretching knocked back most indigenous microbes. Then the cheese sat in a warm space permeated with the smell and presence of yeast. Air- or milk-borne yeasts established, but since the humidity was not kept high, the cheese dried out and a rind could not proceed to grow. The whole room smelled like the cheese and was likely harboring the yeasts that lent this cheese its distinctive mellow, fungal personality.

A house like this *is* the culture for the cheese. The cheese grows out of the house as an extension of the culture of the people, and their food, which is their identity. The house, the people, and the cheese merge as one organism whose feet and hands are the cows, networking their trails through the land as they trim the trees and feed the soil. Communities of microbial magicians and their human helpers connect these symbionts through an invisible mycelium to the rind of the earth, the ultimate host organism.

Once the dried hanging strings were deemed ready, they were cut into more manageable lengths and placed in a bowl. Semi-sour cream was poured over them and folded in until it coated every strand, like pesto on angel-hair pasta. This was left to soak for an hour while some of the cream got absorbed by the strands as they rehydrated. A bit more cream was stirred in to saturate it, and then the sumptuous mixture was packed into small, wide-mouth terra-cotta vessels and placed back in the cold room. The Tenili was also sold in plastic tubs, since there weren't enough pottery ones, and the latter were sold only to locals who would return the handmade vessels, as if they were glass milk bottles. I was told that, in the past, this cheese could be stored long-term by turning the vessels upside down on a layer of ash.

As you can imagine, the cheese was flavorful and gratifying. Tenili is texturally pleasing, as varied consistencies of chewy, gloopy cheese noodles get soggy in God's personal stash of sour cream. I recalled the meals in Norway, where firm skim milk cheeses like pultost and skjørost are mixed in with sour cream. This taking apart of milk, fermenting its pieces separately, then recombining them on the plate in a playful, mouth-pleasing way helps to make a dairy-focused diet less monotonous. It seems to be a common approach in pre-rennet cheese traditions. I was told Tenili was at one time made without rennet, making it a member of an obscure family of rennetless, skim milk, stretched-curd cheeses that are still made in Turkey, Armenia, Iran, and probably other countries between the Mediterranean and India.

There was a depth beyond cream and fresh mozzarella, a definite multilayered yeastiness, and some mystery magic hiding behind the mouth coating of creamy, fatty goodness. *What is that?* I asked myself with furrowed eyebrows. *What am I detecting? What does it remind me of?* Fragments from a memory whispered behind me and were gone as soon as I turned to look at them, the way a dream dissipates from your mind after you wake but leaves a trace of the other side lingering in the room.

The Layers of Terroir

For breakfast, an array of milk foods was waiting for me on the dining table, although no one was in the kitchen. This array included *matsoni*, a yogurt found in Georgia and Armenia that is a bit thin and often has a pleasant, yeasty dimension reminiscent of kefir alongside its light acidity. A mound of butter (also yeasty), which I found too sour, with the delicate aromas typical of grass-fed butter annulled by acid. The white cheese was pretty standard: heavy salt, funky. A bowl of lightly sour cream was heavenly, and had a concentration of a particular aromatic profile that I will attempt to describe: Just walking into the home, an aromatic cloud swallowed me and seemed to permeate furniture and clothing. It smelled like a creamery, like sweet and sour milk, plus a bundle of aromas that formed a common, strong signature characteristic of all the dairy foods served here. It combined a sweet, malty aroma I detect in many naturally fermented cheeses with multiple layers of yeasts, a sulfurous minerality, and an exaggerated aroma of cow milk. It was almost cloyingly sweet, nearly too rich, luxurious, and extravagant. Milk amplified into the realm of decadence.

A glass of boiled milk highlighted the pre-fermentation aspects of the deep terroir manifested on that breakfast table. It instantly reminded me of the yak milk and cream I had in Mongolia, and the milk I was served at the Dragu guesthouse in Kelmend. These milks were rich and eggy. This milk tasted like it had been enriched with egg yolks, extra cream, and minerals. It also had a strong presence of the animal, like grass-fed beef, leather, the way your hand smells after petting a cow. This is the flavor of milk from heritage-breed cows who have diverse diets, which could include legumes, tree leaves, brambles, forbs, herbs, and other plants. They produce less milk but it has higher levels of fat, protein, and minerals. Rather than a particular breed having its own flavor, the heritage breeds just seem to produce a more concentrated, flavorful milk. It's a case of less is more. Less milk, more flavor, higher nutrient density, richer. Quality over quantity. Combine this with a self-selected grain-free diet and the ability of older breeds with generational continuity on a landscape to eat a range of plants in a more goatlike browsing approach, and you have a rich bovine flavor profile. This brings us to two primary layers of potential terroir in cheese: livestock (the breed, herd size, and health) and landscape (their feed and its mineral content, water, the air they breathe).

A third layer is microbial, how fermentation alters and adds to the livestock and landscape layers. The deep terroir I describe really comes into its own as the milk ferments and is transformed into various foods. As with most of the cheeses I have discussed, no tangible starter is added to the milk when making Tenili. But that doesn't mean it doesn't have a culture, a community of beneficial microbes that are primary agents of fermentation. There was likely such a SCOBY being carried forward by these cheesemakers, unintentionally, as a result of their daily use of the same milking equipment, storing milk in the same buckets, turning it into cheese with the same tools. All these materials could have been carriers of cultures that were being fed with fresh milk daily. I think that whole house was a conduit of a base culture that was expressed in all these foods in slightly different ways. I wouldn't call it a starter, because the linear implications of that word miss the mark. The fermentation of the milk in this home was circular. The culture had no beginning, and would only end when the milk stopped flowing. The culture was born from milk, which was never separate from the world around it, so airborne microbes from the environment, and maybe even some from our bodies, could join the band. In this space—the vessel, which is the house—microbial and human cultures overlapped, coexisted, were never separated. This SCOBY that included humans just had to be fed and kept happy, like any community.

The fourth layer of terroir is what I emphasize the most, as it synthesizes and gives form to the first three. Human cultural practices are the most important factor in how milk turns into a specific cheese. They allow the potential of the livestock, land, milk, and its microbes to take form. We create the vessels in which the earth can speak through the goddess nectar of milk, made of sunshine and rocks digested by flowers and vines. The milk of this herd could have been turned into a hundred different cheeses, following traditions from anywhere in the world, but the humans here were making the cheese of their people. The one they learned, that fit into their cuisine, stories, and lives. The livestock here were a result of the community's breeding practices, their relationship with these animals, their rootedness in this place. The landscape had been shaped by the present and past human communities who had altered it, farmed on it, cared for or degraded it. Humans had built the spaces where the animals lived and were milked, where cheese was made and aged, which also housed and shaped microbial communities. The microbes that inhabit milk and become cultures in such spaces are swayed by human touch, by all the decisions and conditions applied by the cheesemakers, which steer a vast microbial potential down the roads that lead to specific cheeses.

Iia and Ruslan didn't send in their milk or cheese for gene sequencing. They didn't identify the specific microbes that were present. They didn't know the Latin names of the various LABs, yeasts, and other microbes that they were cultivating. They didn't need to. They didn't know the exact fat-to-protein ratio of the milk, but they could sense if it was more or less rich, by taste. They didn't know the pH of the cheese but could monitor acidity with the nose, mouth, and eyes. What they did know intimately is what the milk and cheese were supposed to look, feel, and smell like as they fermented properly. They knew the steps of the process, how to use and clean the tools, and why they should follow the steps of this process like a ritual, or when it was necessary to deviate. The recipe for Tenili was not just in their heads but in their whole bodies. To this day, the recipe lives on in a cultural body, the amassed multigenerational experiential knowledge passed down by doing the work and honing the senses to the standards set by tradition. It is this knowledge—combined with the tools, materials, and techniques—that forms the layer of cultural terroir.

The people *are* the recipe.

— CHAPTER TWELVE —

The Flower of Milk

GRAN CANARIA, SPAIN

The boundaries which divide Life from Death are at best shadowy and vague. Who shall say where one ends, and where the other begins?

—Edgar Allan Poe, "The Premature Burial"

Floating on my back in the gentle, lukewarm Atlantic, gazing up as a November sunrise made peaches and nectarines of the clouds, I sighed while counting my blessings and laughing from my belly up. By following an urge to find rare cheeses and to learn from the people making them, I had ended up in the Canary Islands. Part of the Spanish Empire, this volcanic chain off the coast of Morocco is at the edge of the old maps, near the realm of sea monsters and leviathans of the deep unknown, about as far from continental Europe as you can get and still technically be in the European Union. This trip had started over a year ago in Italy, and by sharing photographs and written words, speaking from my heart, and trusting the journey, I found opportunities popping up like dandelions after a spring rain. What began as me rambling around on a shoestring budget with a greasy backpack—sleeping in barns, tents, parks, and airports—had grown into a project that I was finding ways to fund. My first article would be published soon, I had opened avenues for people to donate money, and I was pouring my thoughts into writing longer pieces on Substack. Something had shifted during this year. Instead of saying, "I should have been a writer," I started telling myself and others, "I am a writer." After flipping the script, the vision quickly became reality.

I would spend the final three months before returning to the States doing research for a scholarship called the Daphne Zepos Teaching Award (DZTA). I was granted the award to visit places in Spain where rennet is produced on a small scale, document how it is made, and learn how the choice of rennet impacts a finished cheese. With the rise of FPC (the 100 percent chymosin rennet that is globally dominant), an appreciation for how rennet can play a defining role in a cheese has been lost. Rennet has been reduced to a single enzyme (chymosin) with one task: to coagulate milk. Like every ingredient in modern industrial cheese, FPC is rennet taken as far as possible from the unpredictability of living systems. It is rennet without provenance or ties to the raising and killing of livestock, its variability and complexity stripped away in the name of consistency and control. Cheesemakers often don't know where their rennet comes from, or how they can use different rennets as tools in their craft. The goal of my research was to describe how locally made rennets are a defining trait for certain cheeses, and to share the stories of the holders of this knowledge—people who raise livestock and make their own rennet. These people know exactly where their rennet comes from and how important it is to their cheese. A photographer friend named Alexander "Al" Pomper joined me for this leg, and after spending a few weeks on mainland Spain, we flew to Gran Canaria, the largest and most populated of the Canary Islands.

The islands have been a hub of trade and multiculturalism since serving as a major stopping point in transatlantic voyages during Spain's colonization of the Americas. A plurality of ethnicities can be seen here, with cultural connections to the Caribbean, Venezuela, Argentina, North Africa, Portugal, and southern Spain. This is reflected in the cuisine, which incorporates seafood, bananas, mangos, avocados, local potatoes, and wine, all held together by *mojo*, a family of vibrant sauces based on chiles, garlic, and vinegar. The south side of Gran Canaria is dry, desert in places, and goats are common. The northeast side of the island is exposed to storms coming down from the North Atlantic; heavy winter rains lead to lush, green hillsides in spring. Sheep are the main dairy animal seen on this wetter side. The hills and ravines descending this flank of the island's central volcanic peak are home to a PDO cheese called *Flor de Guía* (flower of Guía, the town from which the cheese is exported) that is made with thistle rennet. It is usually a mixed-milk cheese, with the regulations stipulating that at least 60 percent of the milk come from Canarian sheep, a breed of the island. The plant rennet causes it to ripen quickly, with the interior paste becoming soft and unctuous. It is eaten between twenty-five and forty days old. If aged longer, it can become too bitter. A second cheese called *media flor* (half flower) is made following the same recipe but with a blend of thistle and animal rennet that is aged longer. A third cheese called *Queso de Guía* uses only animal rennet and is aged for six months or longer into a harder cheese that can be grated. Eating these side by side offers a unique insight into how rennet type shapes a cheese. The people here feel it makes such a difference that they give the cheeses different names and ripen them for different lengths of time, understanding how plant and animal rennets differ in their effects.

Al and I left the only city on the island before dawn, driving our rental car out to a quiet beach town where we swam in public pools that caught and held the seawater from high tide. Letting the briny water dry on my skin in the winter sun, I felt like a fresh cheese, ready to ripen into something funky. We drove up winding roads, through otherworldly volcanic landscapes and lush valleys exhibiting an esoteric combination of cactus, palms, succulents, and African plants I had never encountered. Strange birds chirped in the warm and humid island breeze as we crawled up an arroyo carved deep in the basalt to an abandoned hotel. The building was covered in graffiti, which was rapidly being reclaimed by vegetation, folded back into the earth.

The tiny hill towns around Moya are the heart of the geographic zone in which PDO Flor de Guía is made. Agrotourism is developing around the cheese, with a museum detailing the sheep dairying and cheesemaking traditions of the area and a map of producers who offer tastings and tours. We met up with our hosts, Laura and Rocio, who run a cheese shop called Aragüeme, and started down an ill-maintained dirt road, passing orchards of fruit trees on the terraced hillsides. We were heading to visit a *quesería* (Spanish for "creamery") called Altos de Moya to begin learning about the cheese trinity of the island by tasting it from the hands that make it—the hands that hold the seeds of this cheese, and keep them alive with regular replanting.

The Quesos of Moya

Paca Moreno Mendoza greeted us outside the charming white house where she lived with her family at the bottom of a gorge tucked against steep hillsides. It was a tranquil location below the sun-scorched flanks of the volcano, a green oasis with gardens and berry patches. Paca brought us coffee and a cutting board covered in generous wedges of her cheeses. Her face was etched with time and sun, and she didn't smile much. She seemed as stubborn and resistant as a thistle holding its ground in a cow pen. Her demeanor was to the point, as she took time out of her busy day to show us around.

This was my first time tasting Flor de Guía, the pure thistle-rennet cheese made in a pudgy four-and-a-half-pound wheel. Inside is a spreadable, gooey paste that threatens to transgress the border of a thin, rust-yellow rind. When young, it looks similar to smear-ripened cheeses that are washed in salt water. Those orange, sticky, stinky cheeses soften with time due to proteolysis (breakdown of protein) from the bacterial rind community. Flor de Guía has a similar softening of the paste, but it is the action of the thistle rennet that causes a rapid proteolysis. This can be too rapid, creating a cheese with a short shelf life, whose ripening may be erratic.

Paca's young Flor de Guía was herbal and grassy, with a distinct but pleasant bitterness and artichoke flavor from the thistle rennet. An older wheel that we bought from her had a more pronounced bitterness and was breaking down into bold, earthy flavors. It was too much, but tasting it informed my understanding of what happens when the cheese goes over the edge, and why it is consumed in such a specific time frame. There were many gas bubbles in it, which appears to be a general trait of Flor de Guía. The softening of the paste causes the wheels to flatten as they age and the rind can easily break, releasing the gooey, brie-like interior. The short shelf life and fragility mean the cheese is not geared for mass markets and doesn't often leave the island.

The second cheese on Paca's plate was media flor, which uses half thistle and half animal rennet. Paca explained that this combination leads to a more stable cheese with a firmer curd and higher yield. The addition of animal rennet also restrains the accelerated ripening induced by thistle, allowing media flor to be aged for two to six months. The cheese doesn't soften the way Flor de Guía does; the wheels develop flavor more slowly and hold their shape. Paca's media flor had a lighter touch of the bitter, earthy, artichoke, and vegetal aromas that are the trademark stamp of thistle rennet at high dosage. These aromas didn't crush the pleasant and familiar backbone of an aged sheep cheese with a butterscotch and tropical fruit bouquet. There were many layers to this evocative queso, and a mouth-filling balance among sweet, sour, bitter, salty, umami. I was left ready for the next bite, and the next.

Flor de Guía reveals that using pure thistle rennet can cause rapid flavor development and textural softening. Media flor displays the depth that can be achieved by blending. By mixing animal and plant rennet in different ratios, you can alter the impact of the volatile plant enzymes, turning the volume of their voices up or down. Experimenting with blending is moving in the opposite direction from FPC. It is working to increase the potential enzymatic diversity in rennet, using it as a tool that does more than just coagulate milk. It's a way of adding complexity and biological richness back into the process. Rennet is an ingredient whose composition and sourcing should be taken as seriously as those of milk, culture, and salt. In the case of media flor, a cocktail of enzymes is mixed to build a distinct, flavor-packed cheese, with locally sourced rennets that add thick layers of terroir to this cheese.

The third cheese, Queso de Guía, was a rugged and hard aged sheep cheese made with just animal rennet. Like the thistle rennet, this animal rennet was made by the makers until quite recently, with some continuing to do so in a hush-hush way. It gives the cheese attributes of an aged sheep cheese made with whole-stomach rennets; it has some of the spicy mouth-tingling lipase character seen in southern Italian pecorinos. While I enjoyed the bold sharpness induced by the rustic animal rennet, and the novelty and explosiveness of the pure thistle, it was the combination of both in Paca's media flor that impressed me the most. It was such a complex flavor and texture synthesis, with the rough edges and power of each rennet softened and brought into a harmonious whole.

Making Thistle Rennet

Paca had a lot of knowledge about thistle and how to collect it, so I peppered her with questions. Growing up on the island, she had participated in dairying and cheesemaking with her parents and grandparents, who collected thistle flower every summer. Two varieties of thistle are used to make Flor de Guía. *Cynara cardunculus* var. *ferocissima* is a variety of the cardoon thistle that is popular in Morocco. The second type is *C. c.* var. *scolymus*, which is the globe artichoke. It is a cardoon that has been bred to have a huge flower (the artichoke) with few of the spiny spikes that thistles brandish. The stems of multiple cardoon varieties are eaten around the Mediterranean, and the plant has medicinal uses. Paca and her grandchildren collect flowers from the cardoons that grow semi-cultivated around her home. She keeps an eye on them as the flower buds develop in early summer. Once they blossom and reveal showy purple stamens, she cuts off whole flowers and leaves them on a screen to dry. She then laboriously removes the purple stamens, which contain the high concentrations of enzymes necessary to coagulate milk. She said that ten years ago everyone made their own *cuajo de cardo* (cardoon rennet), but increasingly people were purchasing it in bottles from rennet companies on the mainland. She lamented the loss of this knowledge from the skill set of the maker and felt a defining aspect of the cheese was being abandoned.

As the connections between cheesemakers and their primary ingredients are severed, the craft and its output suffer. The bond between the cheese and the place it is made weakens. The move from making their own rennet outside state sanctions according to orally transmitted intergenerational knowledge, to buying it in a bottle from another continent, doesn't sound like a drastic change. It's seen as progress, as one less job to do, as the cheaper, cleaner, more modern way. I seek out and celebrate people like Paca, who understand the importance of having the primary ingredients come through their hands, from the land, the herds, the farm, the home. Having that connection is not necessarily about having control over the process. Rather, it allows us to maintain the web of relationships that can elevate a cheese to the level of a cultural emblem, a symbol of a people's values and their rootedness in a place.

When I meet people like Paca, my faith in the future surges, because she's teaching her grandchildren when to pick the purple flowers that grow on a landscape whose keystone species is sheep. I look up to people like Paca, who understand and lament what is being lost but don't give up in the face of the relentless march of the rebranded inquisition. Many people like Paca are out there, whose lives and cheeses are statements of resistance, a stand taken against mediocrity, greed, and a food system that gives us cheap and plentiful poison while using up the planet as fast as possible. She is the human heart that beats ceaselessly and refuses to be intimidated by the onslaught of modernity's insatiable need to corral life, to make it bend to the whims of colonizer ambitions masquerading as science and commerce.

That heartbeat is what makes me believe in the ability of the human race to overcome our destructive habits, to survive the experiment of the last ten thousand years. It will all be over in the blink of an eye, and we can go back to being humans again, made stronger by having weathered the storm of civilization. Hearts worn and tired, but still beating, still pumping blood made of stars, soil, and seawater.

The drums of dark forces and their obedient armies surround us, but our hearts beat louder, refusing to be silenced by the rage and terror that attempt to traumatize our noble nature into submission. I've been humbled and uplifted by the essential decency at the core of all humanity, which cannot be crushed by the boots of fragile, paranoid empires run by wounded children. The drums of war beat loud, but our hearts beat louder.

Paca's expertise on how to work with thistle rennet is also indicative of the knowledge of plants that is a hallmark of pastoral cultures. Like the sacrificed young animals, the plant is not killed just for its rennet, but also to be eaten and used medicinally. Many plants have enzymes that will coagulate milk, and they often are both medicinal and toxic, depending on dosage and other factors. It seems some of the difficulty of working with plant rennets is the loss of the traditional ecological knowledge that underlies the process. Plants vary in terms of their enzymatic concentration depending on soil type, variety chosen, season, and many other factors. Plant rennets seem to work best when the enzymes are highly concentrated, requiring the plants be harvested and preserved properly. Homemade thistle rennet works for Paca predictably because she has generational wisdom shoring up the process. She has spent her own life practicing and loving the dance of farming and making cheese in tune with the seasons. It is something her body does as naturally as walking. Paca doesn't make cheese. It blossoms from her body, her mind, her home, her heritage. Cheeses seem to sprout from her hands the way fruits grow on trees.

There is an inviting, circular holism to the fact that the enzymes needed to coagulate milk can be found in plants living on the same land that feeds the animals. The rennet grows where the milk grows. Thistles are notorious for thriving defiantly in heavily grazed areas and can pop up like nettles in soil heavily compacted by hoof and machine, where they are one of the few plants livestock generally stay away from. Thistles actually work to remediate soil, by pulling up nutrients with their long taproots while defending themselves and the soil against excessive herbivory. By using thistle rennet, cheesemakers bring this local ecology and the power of this plant into the cheese. The potential for botanical terroir takes a more direct route when we start putting plant extracts directly into milk.

Since the sheep-milking season had not begun, Paca demonstrated the process of making Flor de Guía with milk from her two cows. She used no starter, only a handful of salt. When asked about this early salting, she replied, "That's how my mom and grandmother did it, so I do the same." A small bowlful of purple-brown thistle stamens that had soaked in water overnight sat on a table in the quesería. She scooped out a handful and squeezed, releasing a brown liquid. This was the rennet, and she poured it through a cloth filter, the tea-colored infusion turning the milk a caramel hue as she stirred it in and finished by making the sign of the cross with her spoon. Twenty minutes later, the curd was firm, and she started the cut by again making the sign of the cross, blessing the milk. She used her hand to break the coagulum up gently, then immediately scooped the curds and whey into a plastic form lined with thick cheesecloth. Once it was full, she applied light hand pressure to force the whey to drain out faster, then fit the rest of the curd on top. The cloth was folded over the top to keep the curd in, and she pressed again before flipping the draining curd that hadn't quite become a cohesive wheel. She slowly added more pressure, with both hands distributing the weight she applied by pressing down with her elbows locked.

Paca took Al and me to see her aging space, a temperature-controlled cave underneath her house, carved into the hillside. These caves are common on the island; whole homes have been carved out of the soft rock. Her cheeses were aging on racks made from canes of *Arundo donax*, a plant resembling bamboo that is native to the islands and considered highly invasive in the United States, where it thrives in waterways. One person's invasive species is another's building material. Flor de Guías are soft and fragile in their early days and would pancake out if not wrapped in a band of cloth. Paca's older media flors had mottled, rustic rinds covered in white mold with other colors on top, and many were dry and cracked. These were last season's cheeses, the remnants that she hadn't sold, the imperfect, slightly off cheeses that her family would eat. Around the world, frugal cheesemakers are not eating their best creations but salvaging these less-than-ideal seconds.

The Sheep of the Volcano

A discussion of these cheeses would be incomplete without a look at the sheep and where they live. Farther up the mountain, Al and I spent a few mornings hanging out with a well-known farmer and cheesemaker named Cristóbal Moreno Diaz at his sheep farm, dairy, and quesería: El Cortijo de Caideros. Cristo reminded me of a corn-fed Midwest farmer; he was as strong and stout as an ox. With a shaved bald head and short mustache, he could have been intimidating, but his eyes twinkled like a teddy bear's. The farm sits on a promontory overlooking pastures set on steep cliffs perched above the silent ocean far below, which placidly curves away into infinity. It was the tail end of lambing season, and the new babies were left with their moms overnight. We arrived just as the sun was rising and enjoyed the changing light as the ewes got their lambs ready for school in a rocky night pen. A cacophony of moms and babies talking over one another was accompanied by the chaotic clanging of bells as Cristo released his flock from their pen, sending them running down to the milking parlor with a herding dog snapping at their heels. After milking, the lambs reunited with the ewes and headed out onto steep pastures to graze and take naps in the grandfather sun, which smiled and shone soothing lullaby rays on their sleepy little heads.

The aroma of sheep stirred something in my body that felt deeper than my own life. It called up some ancestral memory or archetypal resonance of a pastoral life, of caring for and living amid streams of woolly waves, rippling up and down ridges and hills, on our way back to winter camp by the seaside, where the flock can feed on salty flats fringed by marshes and the nights aren't so cold. This comforting echo often rises from the collective unconscious into my waking life when I can let myself be immersed in the act of walking as part of a herd, eating the land with them, following their trails carved by their movement. Nothing feels as real and comforting as the smell of sheep as they transform the medicine of the land into food, with the modest power to mend broken agreements.

Canarian sheep are descendants of breeds brought by the Spanish from the mainland, including churro and merino. Like the people of the islands, the sheep are a multiethnic blend. They still possess a high degree of diversity in color and wool quality, with some having very nice fleeces reflecting the merino genetics. Many of them have brown and black splotches and eye patches set within a white or cream background. We sat on the windy hillside, watching as a ewe gave birth to a lamb, which she immediately began cleaning as it lay on the grass, laboriously breathing in its new life. Another set of legs stuck out of the ewe, and we reported this to Cristóbal as he busied about doing the same chores farmers do everywhere. He shrugged and went back to work, having faith in his sheep and their mothering abilities. The hands-off approach that these older breeds allow can shock people accustomed to working with the less independent, less resilient "improved" breeds of dairy sheep. These Canarian sheep are survivors; keeping their young alive is what they do. All around us, in an idyllic pastoral scene, attentive mothers watched over their babes. Life did what it does. The ocean below moved in cyclical currents. The sun rose as effortlessly as a sprout breaking its seed walls. The lamb dropped from its mother's body into this world as the wind shook a leaf loose from a tree.

The milk at this point in the year was mainly left for the young, who were being fattened up for sale as Christmas lamb. The intimate connection between milk and meat, life and death, is not shied away from in places like this, where a culture is firmly grounded in shepherding and cheesemaking. Cristóbal adores his flock, fights for the sheep of his island, and calmly faces the reality that young sheep need to be sacrificed so the milk they would have drunk can become the cheese that he also loves and identifies with. There is no separation here between farming and cheesemaking. The cheese grows out of the farm and is intimately connected with the sacrifice of young animals. Even if it is made from a plant instead of an organ, the cheese implies the death of animals, which is integral to dairying. For death is as integral to life as night is to day.

— CHAPTER THIRTEEN —

Rennet Strongholds

LA PALMA, SPAIN

We stand for what we stand on.

—Wendell Berry

From Gran Canaria, a short flight on a small plane took Al and me to La Palma on the west end of the volcanic Canary Island chain. Its steep canyon-carved sides rise quickly to high mountains crowned with calderas and cinder cones. La Palma is a work in progress, as it continues to grow from active volcanic vents. It receives more rain than the other islands, earning it the title *Isla Verde.* The green island has gone through a series of monocrop boom-busts, including tobacco, wine grapes, and now bananas. What has lasted through these waves is a persistent pastoral cheese tradition rooted in the pre-European Guanche culture. It is one of Europe's last remaining strongholds of small-scale rennet making. All of the cheesemakers we visited on La Palma raise the goat breed of the island and turn its milk into cheese with rennet made from their own herds. Many of them use sea salt that is evaporated here as well, leading to another element of terroir in their cheeses. The enforcement of food safety laws has recently begun, resulting in the slow erosion of this dairy sovereignty as commercial starters and rennet are being adopted. The makers here are well organized and fighting to fit their traditional practices into this legal framework. What they have accomplished is a success story that can be replicated elsewhere.

The rennet strongholds I have visited share some telling traits. They are usually on the fringes of nations, in mountainous borderlands (the Basque Country, Tusheti), on islands (Sardinia) or, in the case of La Palma, on an island at the periphery of Europe. There is always a cultural connection to traditions of pastoral dairying, and a sense of pride in a cheese that serves as an emblem of this lifeway. A conservative insistence on using old tools, following the recipe without deviation, and preparing the ingredients yourself are hallmarks of these rare places. A local breed of sheep or goat is kept and revered as a crucial part of the cultural identity, the cheese, and the life of a people. These animals are loved, and efforts are made to prevent the genetic dilution of the breed. The esteemed breed is also a source of meat, and this reality is not hidden or seen as an ethical dilemma. The connection between dairying and eating meat remains strong and is often seen as a part of maintaining culinary and agricultural traditions. The food, land, agriculture, culture, and people themselves are all inextricably one thing. This cultural complex has been defended for centuries or has somehow survived despite physical invasion by various armies and empires.

In rennet strongholds, the eating of the young males from dairy herds is celebrated as an act of cultural vitality, often associated with holidays or festivals during or shortly after the time of year when animals would naturally be giving birth, when grasslands green up and there is abundant food. There is a vestige of this in many places where dairy sovereignty has been lost, such as in the eating of lamb at Easter or Christmas. It is likely that the consumption of very young ruminants (lamb, kid, veal) is tied to the historical reality of a winter or spring sacrifice of young male dairy animals at the start of the cheesemaking season, which would have provided the rennet to be used that year. This practice continues in these strongholds, where despite the time, cost, and hassle involved, people view the use of this rennet as integral to their cheese and as an act of opposition to the corporate food system, which is rightfully seen as the same old imperial game, with new logos and a better marketing strategy.

In much of the world, the role of women as cheesemakers was lost as the craft became a male-dominated industry, a transition that started with the Industrial Revolution and was fairly complete by the mid-twentieth century. I've been places where the women were the ones making the cheese, but their husbands or sons seemed to take credit for it and were the public faces of the business. In many overtly or subtly patriarchal societies, I had to go through the men to see the women make cheese. The men answered my questions, even when they were obviously not immersed in the practice. This was not the case on La Palma, where the small cheese companies I was interested in were all run by women. I was told to contact them. When I arrived, they greeted me and showed me around, clearly running the show. There were male cheesemakers here too, and husband-wife teams seemed to be the norm. Often I found that the husbands managed the herds, but I got the impression that cheesemaking was still a part of the feminine domain on the island—another indication of the survival of the old ways here.

These themes I keep touching on—of motherhood, birth, death, sacrifice, and how we treat the land—are dear to my heart, and interconnected. We are all participants in an abusive relationship, where we are both abuser and abused. The way we treat livestock is directly correlated to how we treat the land. The treatment of women and mothers in our society reflects our views toward the environment, the planet, the wild, and the domesticated. Abuse and subjugation of the feminine is a part of our denial of the sanctity of life, of viewing land as something that can be owned, plowed, seeded, forced to bear fruit, fenced in, exorcised of wild animals, weeds, and unrestrained biology. Our efforts to enslave livestock, motherhood, plants, and life in general have led to ecological destruction and a broken food system. This sadistic relationship is not a reflection of who we inherently are. We have just been in this abusive relationship for long enough that it feels normal; we've internalized the twisted logic of our self-oppression, and we come up with all sorts of stories to justify it rather than walk away or confront the truth. Our seemingly flawed natures are actually a temporary deviation based on an unhealthy story about our role in the world, a broken contract with the holy. What needs to be reclaimed, so the contract and connection can be restored, is a sacred understanding of life. We need a new story, or to revive some of the older ones that once served us well.

Paula and Her Goats

Our first visit was with Paula Pérez Rodríguez, a young woman who had recently started a cheese company called Granja Los Lirios working with the milk of her small herd of Cabra Palmera (La Palma goats). She clearly adored her herd, which, she said, "act as though I am their mother." Paula was bubbly and petite, handling the heavy workload of operating a farmstead cheese business with a grace and maturity remarkable for someone in their twenties. She took us to meet the cantankerous goats as they clambered on rocky cliffs close to a barn and milking parlor, overlooking the wide ocean.

Displaying a range of colors and hairstyles, there are a number of distinct lines of Cabra Palmera named for visual distinctions, and various lineages kept by certain families or found in particular districts. Some don a brown mohawk along their spine and shaggy blond mullets down their back legs. Others have a curly red bouffant or parted black bangs above their forehead. When I asked them who their stylist was, they just scoffed and resumed knocking heads. Their horns grow out to the sides then curve up abruptly, starting to spiral as they get older. The Cabra Palmera descend from goats raised by the indigenous people of the islands, the Guanches, and predate European contact. Breeds from other islands have been mixed in, adding genetics from Europe. Efforts have been made to select back to what is considered the true breed. Few places have done such a good job of maintaining their heritage breed as a working animal, not just as a sideshow. The Cabra Palmera display a wildness I have only seen in one other breed of goat, the Girgentana from Sicily. The two breeds share a skittishness and loaded-spring energy and are visibly apprehensive of outsiders coming onto their turf. They express normal goat behavior, but exaggerated. More goaty than goat, or what goats inherently are, when humans allow them to be. I've also heard rumors that neither of these breeds tolerates weak or wounded members in the herd, and will sometimes attack or kill them, following a kind of culling instinct to keep the herd safe and strong.

Paula told us she started her operation because she wants to make and keep alive the cheeses her grandmother fed her as a child. Grandma is the highest force on the planet—not my grandmother or yours, but grandmotherness as a universal principle. While we chatted, she unloaded a bunch of vegetation from her truck, including tall grasses and a native legume called *tagasaste* that looks like alfalfa and can reach tree heights. A major feed is green bananas gleaned from the local industry, which are chopped up and fed with skin on. We saw similar stockpiles of freshly cut fodder at every farm we visited. There is some extensive grazing still happening, but this is disappearing as farmers become more reliant on alfalfa grown on mainland Spain and shipped to the islands as bales or pellets. We smiled and laughed as Paula lovingly scolded and joked with the impudent mob of animals rushing her for snacks. Paula effortlessly embodied in her actions and presence the notion that rearing livestock and making cheese are intimately linked to birth, child-rearing, and the interweaving of feminine and masculine principles. She was a part of the herd; they were family rather than animals that she owned. When I run into young people like her who are deciding to push back against the loss of the commonsense wisdom contained in agrarian pastoral lifeways, it fills me with hope. Hope and optimism, because I see that more people are choosing to take a stand for the land, for the wild, for the ground that feeds us all. For the earth mother whom we are slowly turning toward as we grope in the dark, searching for our way back home.

The wisdom we need cannot come from a book, lecture, or political party. It seeps up from the ground, is found in wild plants, and is heard in the call of a baby goat seeking its mother. It's a timeless wisdom shared through eyes and hearts, through new languages created in chaotic times, through healing the amnesia of the present as we break open the futures that we can step into. Before these new futures can manifest, they must be spoken. I see more and more of us standing up and speaking truths that are eternal, that can be said in any language, even conveyed through silent action, the motions of our bodies, the expressions on our faces. Paula and her goats, ungovernable, clinging stubbornly to a fiery, unstable mountain jutting from the sea, are the reason we will survive.

Cheese Island

The cheese of the island is called Queso Palmero and has a PDO that requires the use of raw milk from certified herds of Cabra Palmera. This insistence on a single breed whose milk is never stored cold but made into cheese immediately after milking is rare in the world of protected names, which often makes compromises to allow for large-scale production. There is a wide diversity of cheeses within this name, mainly made by families who operate queserías attached to or near their own homes. Queso Palmero has never been, and probably never will be, made on an industrial scale and marketed aggressively. Like so many of the cheeses that I have spoken of in this book, the cheese of La Palma will never reach a shop near you. Most of it is consumed on the islands. The PDO cheeses have a black label that includes the name of the producer. A separate green tag is used for cheeses made from herds that are on pasture, and that are bred seasonally rather than year-round. As skeptical as I am about protected names, this PDO shows that the approach can actually be used to serve the small-scale producers and help them maintain their time-tested practices, keeping the cheese of their people alive and moving forward, adapting while maintaining its essence.

The cheese is sold at various ages with their own names: *fresco*, *tierno*, *semicurado*, *curado*, and *anejo*. It can be anywhere from a small, soft, mild cheese to a hard and boldly flavored large wheel. Some cheesemakers consider there to be four distinct subtypes based on the material used to smoke the cheese. The various techniques of making rennet add a third variable that is not set in stone, allowing the maker greater latitude in crafting a cheese that expresses their district, farm, or family, or their own creativity. The anarchical nature of Queso Palmero means it is not really one cheese, with one flavor profile. It defies categorization. The beauty of this cheese cannot be reduced to a recipe, to rules, to a uniform product. The goats are resilient and valuable because they lack uniformity. A one-size-fits-all farming system imposed from above can never match the diversity of approaches that have been built slowly, from the ground up, with trial, error, and ongoing experimentation. In the defiance of Queso Palmero, in its refusal to be one commodified thing, lies the transcendental deliciousness that I seek, that shows us what humans can accomplish when we allow the land to speak through our hands, our cultures giving voice to the holy earth. Queso Palmero tastes like a mother's laughter and love for her children. Multifaceted, infinitely deep. Soft and caring yet fierce and serious, prone to fiery eruptions. Queso Palmero is an embodiment of the first rule of life on this planet: Don't mess with Mom.

Smoked Cheese

Queso Palmero is typically but not always smoked using a variety of materials, dependent on local abundance and the style of the maker. The makers I spoke to all considered the material used to be crucial to the flavor of their cheese, and some of them blended materials to achieve the preferred balance. Almond shells, dried prickly pear cactus, the needles and wood of Canary pines, and chestnut were all listed, and their attributes praised or denigrated. Smoke is a not a singular substance; it has a complex chemistry, and distinct esters and phenols can be released from different materials. The cheeses that are destined to be aged longer are sometimes left unsmoked, as are young frescos. The knowledge around smoking cheese is highly advanced on La Palma, and by utilizing local materials and developing micro-regional techniques, smoking becomes an expression of botanical and cultural terroir. On an active volcano, this inclusion of fire and smoke feels even more fitting, another taste of this place.

The origins of smoking cheese on La Palma are unclear, but it's a strategy I often see in humid places where it is difficult to store cheeses in cool cellar conditions (50–55°F, 10–13°C). Keeping cheese at a stable temperature and humidity in a cellar or cave is a rare luxury. Cheeses stored above these temperatures often attract flies, which can lay eggs in a cheese, with subsequent eruptions of maggots: the next generation of flies. This appears to have been a common issue or normal trait of cheese in pre-refrigeration times. In *A Tour Through the Whole Island of Great Britain*, Daniel Defoe offers this observation from the early eighteenth century:

> *We pass'd Stilton, a town famous for cheese, which is call'd our English Parmesan, and is brought to table with the mites and maggots round it, so thick, that they bring a spoon with them for you to eat the mites with, as you do the cheese.*

People sometimes grew enamored with this lively fromage, and I've heard stories in many places that some of the old folks have fond memories of eating maggoty cheese, prizing it as a delicacy. Only tales remain of most of these lacto-insect chimeras, but it became the defining trait of the famous Sardinian casu marzu. Smoked cheeses are less desirable to flies, and the treatment kick-starts a dehydrated rind layer while likely dissuading molds from growing. Smoking cheese has not always been about flavoring or putting liquid smoke into otherwise bland, mass-market, shrink-wrapped blocks. It has long served as first line of defense, as a key player that can be used along with salt to get a cheese heading in the desired direction and keep unwanted microbes or larger life-forms at bay. It is an ancient and incidental practice, as cheese was often made over fire and then hung in the rafters above, giving it a light smoking.

To learn about smoke and rennet, we visited Nazareth Pérez González of Granja Los Barbusanos, which is cocooned in a rugged ravine swathed with lush green vegetation. Her herd wore wide leather collars and antique brass bells, which clangored in alarm as the goats retreated from our presence, two grinning foreigners with backpacks and cameras. They peered out from the most unique and prehistoric sleeping shelters I had ever come across in my travels: caves carved back into porous, volcanic cliffs, large enough to house a few hundred surly, surreal capricorns. We also saw a collection of week-old kids who were crying out to us for milk. They were obviously accustomed to being bottle-fed and considered any humans to be their mothers.

Like any good cheesemaker, Nazareth jumped into multiple tasks upon entering the creamery, while also giving us a tour. She handed us a plastic container that housed two abomasa in a thick, slimy, congealed brine, like a broth with lots of collagen. The stomachs were full of the milk the young goats had drunk as their final meal, which had coagulated into cheese naturally, inside the organ. These were from the previous winter's kids, which were taken to the slaughterhouse on the island, which operates at EU standards, like a USDA-inspected plant in the United States. The fifteen-to-twenty-day-old kids were slaughtered and butchered there, and the meat packaged for the farmers. The abomasa could be made into a paste by the slaughterhouse or given back to the farmer with yellow tags attached to them that have a traceable serial number. This is how the makers of Queso Palmero have found a way to continue making rennet from their herds while complying with food safety law.

Nazareth follows one of the main techniques used on the island. Fresh abomasa are placed in containers and buried in copious amounts of sea salt from the island. The salt pulls moisture, fat, and collagen from the organ and the connective tissue that has been left on. Much of the salt dissolves, creating a gloopy, saturated brine. To turn this into rennet, a whole abomasum including its internal cheese is mixed up in a blender with water and some of the brine. This creates a slurry that is filtered, generating a salty liquid rennet that is stored refrigerated in water bottles for up to two months. Nazareth said one abomasum would coagulate about four hundred gallons of milk. The gang of baby goats we saw outside would become the rennet Nazareth would use to make more cheese, from their mothers' milk. Rennet is provided by the act of dairying, by the herd itself, which gives us so many gifts. I know many people find the situation unethical, but I have witnessed those who still make rennet from their herds displaying the highest degree of love and respect for their animals. In visiting rennet strongholds, I have found highly ethical forms of dairying that acknowledge the sacrifice of living things, which is fundamental to eating and sustaining life. Maybe we shouldn't pity these young sacrificial beings. Maybe we should honor and revere them as the chosen ones, whose lives are completed in a consecrated act that ripples outward in the sacred circles in which our food and lives are enmeshed.

All of the queserías we visited had large metal smokers called *ahumadors* built into a side room with overhead ventilation to pull the smoke out. Most smoked cheeses are cold-smoked, where the smoke has a chance to cool before enveloping the cheese, with which it spends an extended time mingling. Queso Palmero is smoked hot and fast on a metal rack, leaving distinct grill marks on the wheels. Nazareth placed a thick bed of Canary pine needles below the rack, covered it in cheese, and lit a smoldering fire on one end. As smoke began to billow, she placed a large sheet over the cheeses to let the volatile compounds in the smoke infuse them, spraying it with a hose to generate even more smoke. If the fire flared up, she quickly sprayed it out. After ten minutes she lifted the cloth, and if the wheels had a nice golden-brown layer and looked like a properly roasted marshmallow, she pulled them off. She washed off this black-brown layer, a greasy soot grime that sometimes appears on improperly smoked meats. Left behind were the sexy tan lines from the grill. A pleasant and lightly smoked character had been passed to the young cheeses, which was mainly perceivable on the rind.

There are other rind treatments used on the islands that are likely attempts to preserve humidity and/or deter insects, such as coating wheels in a mixture of ground grains, olive oil, and paprika or cumin. I used to consider smoked cheeses and those coated in seasonings to be gimmicky, not real cheese. Now I understand that these treatments have a place in the toolbox and play a functional role in many traditional cheeses. Smoke and coatings can become a literal layer of terroir, another dimension of a culture and its landscape in a cheese.

In countries where farm rennet making has disappeared, cheesemakers often voice that the process feels out of reach, difficult, and unrealistic, if not impossible to achieve in seemingly unavoidable legal frameworks that discourage the practice. What I discovered is that crafting plant and animal rennet is a fairly simple process, and that these rennets can be relied on to consistently coagulate milk. Not only are these local rennets dependable, but they can also introduce nuanced signature flavors that are impossible to achieve with the generic single-enzyme FPC. Making rennet and starter cultures is straightforward, time tested, and safe. The only difficulty lies in having these processes sanctioned by national governments, which seem to have a strong bias against these foundational aspects of dairy sovereignty. If the industrial products that are rapidly replacing locally procured cultures, rennet, and salt are not leading to safer or more delicious cheese, we should stop using them. We should take a serious look at what we can learn from people who never made these compromises, who still hold the seeds of cheese—of both what cheese was and what it can become again. We can take a cue from the last remaining rennet strongholds, from the goats of La Palma, who, along with their human partners, can show us what resistance looks and tastes like, who are taking a stand for what they stand on.

— CHAPTER FOURTEEN —

The Salty Seas of Planet Cheese

TAMIL NADU, INDIA

We are tied to the ocean. And when we go back to the sea, whether it is to sail or to watch—we are going back from whence we came.

—John F. Kennedy

A man wearing a blue checkered cloth around his waist hand-milked a buffalo as the horns and engines of a busy road droned in the background, the white noise of urban India. The day's milk was brought to a bustling kitchen and made immediately into a range of cheeses that would be served at the restaurant called The Farm, in Tamil Nadu, southern India. I sat under the shade of tall bamboo in a garden patio tucked between the buildings of this oasis of greenery that was being swallowed by a rapidly growing city. Sipping sweet buffalo milk chai and tasting the various lactic delights from a cheese plate, I was amazed that even in this setting, far from the clean mountain air and rolling rural grasslands that I typically roam through, a taste of place emerged. Rather than tasting like city streets, the cheeses spoke of tropical fruits and sandy soil cooled by sea breezes. They expressed milk's inherent life-sustaining goodness. They tasted like home. Not my home, but the home of others who welcomed me in. They tasted like hospitality, like a door held open by a smiling face. They tasted like eye contact between two souls who want to meet. They tasted like the aspect of Indian culture that deeply affected me and so many other travelers: an openness of heart.

After sharing the stories and techniques that years of cheese trekking had revealed to me, I was being invited to many places around the world through connections made via social media. The project was shifting from the gathering of information and experiences to disseminating—to seeing how I could benefit cheesemakers who were interested in applying the approaches and philosophy I was writing about. I had begun offering five-day natural cheesemaking workshops in collaboration with host farms through a mobile educational endeavor I started called Sour Milk School. After three months of exploring the pastoral dairying traditions in India and the rise of an innovative cheese movement, I was wrapping up the trip by participating in the first gathering of Indian artisan cheesemakers, held at The Farm.

Arul Futnani and Shalini Philip, the owners and operators of The Farm, had invited me to come stay years prior to my visit. From seeing photos of their cheeses and dishes, and getting a digital feel of the ambience, I could tell it was a place I needed to visit in person, to experience and taste. I reached out on Instagram, talked with Arul on the phone, and decided I would go to southern India someday, then put that vision on the back burner, to let it slowly simmer into a broth to add to life's stew when the timing felt right. Their model of merging dairying, cheesemaking, and a restaurant into a diversified farm, crafting an abundance of value-added products, appealed to me as one way that small-scale dairying could remain viable. When I learned they were also a micro coffee roaster, it kind of sealed the deal. I don't know that there is another restaurant at a buffalo dairy and creamery that roasts coffee.

Arul and Shalini were welcoming hosts, easygoing and fun to be around, obsessive epicureans skilled in combining flavors and putting together plates in a way that was colorful and artful without being ostentatious or over-embellished. We quickly became friends, sitting in an outdoor dining room covered by a broad thatched roof, eating hearty breakfasts of local specialties such as *idli* (fluffy steamed fermented rice and lentil cakes) served with dal, coconut red chile sambal, yogurt, and a rainbow of in-house chutneys. Fiery, sour, vibrantly spiced smorgasbords after casual cups of light-roast pour-over coffee felt climatically appropriate as the warm mornings grew into sticky, hot days.

The cuisines of India are some of the most milk-centric on the planet, and yet only a small range of traditional cheeses can be found. Paneer is the most famous, based on **acid-and-heat coagulation**, a technique popular from Turkey to the Himalayas, and north to Mongolia. A major use of milk is for making butter, which is further concentrated as ghee, with the skim milk being used for a yogurt called dahi. Chai is another main route for whole or skim milk, and you are never far from a crowded stall where vendors whip up batch after batch with theatrical pouring rituals for the streams of customers who crowd the sidewalks, slurping sweet milk tea lightly seasoned with lemongrass or ginger. A whole family of sweets based on condensed, dehydrated milk are consumed and offered to the gods, who are renowned the world over for having a serious sweet tooth. Milk is sacred in Hindu cosmology, being poured as offerings onto Shiva icons, and ghee is burned in temple oil lamps. Milk in India is almost always boiled before being used, and rennet-coagulated cheeses were uncommon until recently.

The Farm sits near the ocean on what was formerly a rural agricultural landscape of sandy flats resting just above lagoons and salt marshes, sprinkled with wetlands and ponds. As the city of Chennai has spread, the area has mushroomed with tech development and lost its farming economy. The Farm survives and thrives by being diversified and selling high-quality products direct to consumers through an on-site farm store. The main attraction is the restaurant, which has a well-earned reputation for excellence and draws a crowd of locals, expats, and tourists.

When I arrived at The Farm, I was not expecting to find a case of deep terroir. A gas station and busy road sat just outside the walls, and the air, thick with humidity, carried aromas of hot pavement, exhaust fumes, burning trash. My cheese snobbery was challenged when I tasted these raw tommes and was met by something complex and seductive that could have come from the mountains of northern Italy rather than the subtropical coast of southern India. At first the cheeses seemed a bit out of place culturally, but eaten on a plate with local jams and fruit, they slowly melted into the cohesion of a new arrangement of transcultural terroir. I sat back in my chair and closed my eyes, sighing heavily as the melancholic bliss of a cheese satori kicked in. The brightly colored birds laughed as they jumped through the hedges, their jovial language joining the symphony of human sounds. A gentle breeze taught the tops of tall palms a contemporary dance. The smell of incense burning from the doorway of a tiny temple tickled my nose.

Cheese styles have coevolved alongside cultures, cuisines, and farming systems born in particular climates. They have also been carried around the world on trade routes like spores on the wind, folded into other cuisines like novel spices. This transculturation represents the flip side of my focus on place-based traditional cheeses and breeds: the innovative transplanting and reinterpreting of techniques, the blending of cultures. When done with respect and intention, experiments in these culinary frontiers feel not colonial but liberatory and point to the potential of cheeses that transcend borders and markets. These cheeses can return us to our bodies, grounding us where we are now, in a disfigured world, shattered and abused by the forces of greed. These cheeses tell us to put our feet and hands back in the dirt, to start making more of it, to right the wrongs, plant the seeds, adapt them to our future, rebuild sane foodways that feed all life, wherever we happen to be. These cheeses bring us to the front of the changing tide, as civilization's destructive high starts to ebb, and we are pulled back into the infinite ocean of our birth, to bathe in the salty tears of our mother's embrace. Just as the sea is pushed by the moon, an invisible hand is pushing the holy seeds of our indigenous human dignity back into the living crust of the earth, back into our role as caretakers, cultivators, and eaters of the sublime rind.

Coastal Terroir

Arul and Shalini described how they had worked from the ground up to introduce aged cheeses with natural rinds and textural flair to Indian palates that were more geared toward fresh cheese, yogurt, and the use of boiled milk in chai and coffee. They started by making mozzarella and feta, styles that just about anyone can appreciate. As they built a trusting customer base, they added in more adventurous offerings with colorful rinds based on styles from Europe not commonly seen in India. For instance, a truncated pyramid with white mold, and two raw milk tommes, one from cow's milk and the other from buffalo. For me, it was in these two smear-rind tommes that the milk and farming practices of The Farm really shone, expressing the terroir of place and people. Of people in place. Of a peopled place.

Many makers of this smear-ripened or washed-rind style add packaged ripening cultures to establish the communities that lead to sticky, stinky, orange-red-pink pigmented rinds. This often includes a yeast and one or more *Brevibacterium*, such as the well-known *B. linens*. People fixate on this single species, but it is generally a community of salt-tolerant, bacterially prominent microbes that form a smear, at times without the presence of *B. linens*, even if it is added to the milk or the 2 to 5 percent washing brine. When rinds are kept moist with regular brine washing in a conducive environment, the smear begins to form. I've seen smear rinds form like this naturally (without adding the ripening cultures) in many humid, coastal regions, where they often seem to be a part of indigenous milk microbiota. This seems quite common along the California coast, which is bathed in fog for much of the summer. It is logical that these salt-loving communities are pronounced near bodies of salt water, especially close to boundaries between sea and land. The line is blurry, as marine fog, wind, and flooding introduce salt to a continent's crust, its rind. In these marine-terrestrial interfaces, salt-tolerant microbes are likely living on pasture plants, hay, and bedding straw, and on the bodies of animals who graze and sleep on them. Marine microbes have been found in analyses of smear-ripened cheeses made far from the sea, with the sea salt used to wash them serving as the source.

The coast of a continent is a biologically rich edge zone, home to a diversity of species where life grows in layers of rapid transition between biotic communities. Salty waves come and go as tides, washing the coast, developing its striated ecosystem and pronounced aromas. The rind of a cheese is its coastline, a rocky shore, a boundary between the creamy land inside and the outside world, a sea of air swimming with critters. The rind, like a beach, is a liminal space where exchanges take place, a permeable border that connects and divides various beings and states of being. Seen through a microscope, cheese rinds look similar to a rainforest or coral reef, with different layers, from soil to underbrush and canopy, and niches being created and occupied by complex communities of organisms. These organisms interact with the geology under their feet and the space in which the cheese ages, its climate. Born from macro ecologies on the earth's thin crust, a cheese is a fractal of the planet, with its own geology, ecologies, biological communities shifting with climate, weather, and the scheming hands of invisible gods.

The buffalo tomme called Queso Coromandel sported a gorgeous sticky-orange smear as a backdrop with white, blue, and gray molds adding their pigments to the foreground of this organic abstract-expressionist-painted planet. The young wheels were wiped down regularly with brine and snuggled together in wooden boxes, which were stacked up inside a chest freezer turned aging space. The sticky smear developed naturally, encouraged to grow by regular saltwater rubdowns and the cool, moist micro-atmosphere inside the boxes. A patina of white mold could be seen flourishing on the boxes, tendrils of mycelium trailing along the grain of the wood. Once the smear had developed, washing stopped and this fungal community that contained wild strains of *Geotrichum candidum* and mucor started to spread onto the cheese, creating a polychromatic fluff layer on top of the orange smear. This diverse rind ecology softened the cheese underneath and added a range of sweet, meaty, bitter, and earthy flavors as the fatty buffalo milk composted into deliciousness.

No commercial ripening cultures were added to get this rind microbiota to form; it sprang from the milk, from spores in the air, and from the biofilms inhabiting the wooden boxes that were a crucial part of this cheese. The living rind was cultivated by simply washing with salt water. The salt was selecting for the smear community, and against other microbes. Once washing stopped, these others could come back in, onto the soil prepared by the smear, in a succession of microbes that could grow into a miniature tropical rainforest. As it grew and engulfed the cheese, it gave a voice to this place, to The Farm, to its climate and landscape, and to how a family had farmed here for decades.

Even when commercial starters are used for the primary fermentation—as was the case with these tommes—terroir can still express when given enough space. It is usually in part a matter of using a lower dose of the purchased cultures so the raw milk microbes can still shine through. At The Farm, the milk had a lively microbiota that was born into a hot, humid climate that felt designed to encourage fermentation. I was there in the coolest part of the year, but it felt like walking around inside a koji incubation chamber. Mold dripped from the clouds. The milk had a strong heartbeat, and when left raw, its fermentation from native microbes was almost unstoppable. Experimenting with generating clabber starters from both cow and buffalo milks showed that we had to either keep the starters in an air-conditioned space or feed them twice a day to keep them from fermenting too far, too fast. That milk wanted to be cheese; it was actually hard to keep it from realizing its destiny. Why would we want to, anyway?

Milk Poetry

Arul took me on a walk around the large yard where the cows and buffalo spent the nights. A man opened a gate to allow an ox to come in, pulling a cart of green plants cut from one of the adjacent fields. Another worker sprayed down the mortared paving stones that made up the yard. I caught a whiff of a familiar aroma and got closer to the freshly washed paving stones. It was the unmistakable aroma of a washed-rind cheese, a pungent combination of overripe fruits, body aroma, and barnyard that was somehow pleasant, a great example of what Michael Pollan calls "the erotics of disgust." These cheeses often contain a whisper of decay, a suggestion of death, and the possibility of milk's glorious afterlife. A washed-rind cheese aging in wood boxes doesn't smell like a dirty cow barn, but there is an aromatic and perhaps microbial connection on some level.

At The Farm, the yard was kept clean enough that the herd could eat their cut fodder directly off the ground. Rather than trying to exorcise life from the process with chemicals, a balance was struck by regimented cleaning utilizing water and ample sunlight, a powerful natural disinfectant. The manure, urine, and wash water ran down a gutter to a ditch on the way to returning nutrients to the land where the food for the herd grew. Lush fields of bananas, tall grasses, and vines were crossed by a grid of irrigation ditches, with the nutrient-charged rinse water gravity-flowing out onto the biological filter of the earth. This rinse water fed the plants that would be given back to sacred bovine goddesses. Manure is a gift that is still appreciated in places where the earth is still embraced as a living entity to be worshipped rather than treated a cadaver to dissect and turn into money. It was a remarkably self-contained, circular process. At night a large tree in the yard served as the roost for dozens of flying fox bats, who could be seen stretching their wings in the bright lights of the gas station on the other side of the wall. As the city squeezed against these walls, more birds, insects, and wildlife appeared on the grounds, a sanctuary for life amid forces that seemed bent on eradicating it. Like the remote villages I had visited in mountains and deserts near nation-state borders, The Farm was also an island of nourishment, an oasis that fed people, animals, and the land. But it was set in the middle of an urban sea of chaos, making its atmosphere of health and calm more pronounced.

The next piece of the puzzle of terroir came when I joined the herders as they took the cows and buffalo out for their daily grazing circuits in the surrounding marshlands. After feeding on the vegetation that grew rampantly in southern India's eternal summer, the buffalo and some of the cows gleefully filed into a large pond to relax, with just the tops of their heads and snouts sticking out of the water like bovine alligators. Acting on an urge, I stuck my finger in the water and tasted it. As I suspected, it was nearly as salty as the sea. The bodies of the animals were also being brine-washed. When the herders prodded them with sticks and words, the buffalo reluctantly trudged back through palm groves, returning home for evening milking, the salty pond water drying on their rinds.

Buffalo and cows wallowed in salt marshes full of little fish, insects, and no doubt a diversity of microbes. The herd ate locally grown fodder and were milked in a porous stone yard. The milk was carried a few hundred feet and made into cheese immediately, with no intervening refrigeration to damage its thriving ecology. The resulting cheeses were washed in salt water, which encouraged a delicate perfume and umami booster that was possibly linked back to those saltwater ponds, to the sea, to the yard in which they loafed. I didn't have direct evidence that this web of connection was in fact leading to a microbial terroir on the rinds of these cheeses. We likely couldn't tease that web apart without breaking it. But reality can also be tasted, my mouth and nose informed me, and those cheeses had *it*, even if we can't quantify, isolate, and publish in peer-reviewed journals just what exactly *it* is. These had the magic depth of flavor, the sparkle of wildness, the look in the eyes of the fox before it silently jumps a stone wall.

The cheeses that I have showcased throughout this book, which have signature terroir, are like poems, unique arrangements of a standard set of words, the lexicon of milk chemistry. Those words can be used in a limited, standardized way, or they can be used creatively, to express an infinite range of thoughts, emotions, and places. Cheeses that voice terroir as milk poetry evoke more than mere sensations in my mouth and nose; they take me beyond the intellectual exercise of naming flavors, comparing this cheese with others. They evoke emotional memories inside me, of working and living on farms, of walking in unfamiliar lands, of meals with new friends, of smiling shepherds. They ground me, bring me home. I can't really offer tasting notes or point toward something you could find in a cheese shop that would be an equivalent. The experience of terroir is incomplete outside the place of its origin—outside its sounds, smells, air, stones, temples, flowers, forests, and people. Oysters taste better when you're near the sea, engulfed in the briny air, hearing the mockery of gulls and the gentle song of the waves washing the earth's rind. The edge is where the sea fruits oysters. If you want oysters, go to the sea. If you want to try the cheeses of The Farm, go to The Farm.

The Cows Come Home

Arul's father had run the buffalo and cow dairy for the last fifty years. He had mixed various breeds following a fairly hands-off approach, letting the bulls do their thing without following a strict plan. This blend of artificial and natural selection reminded me of the breeding programs of many pastoral groups, where humans breed animals intentionally but don't try to control everything. Nature still has a hand in the game. This allows the genetics of a herd to be shaped by natural forces in tandem with the farming system itself. Over a fairly short amount of time, distinct genetic lines had emerged on The Farm. The herd was a mix of breeds and genetics and appeared very healthy. Years of experimentation by families, farmers, and workers can lead to a place-based interspecies collaboration that bears fruit in the form of cheeses whose seeds are worth keeping alive. Rather than the last specimens of fading traditions, what I found at The Farm were the seeds of future cheeses.

Most of us are not able to run away to the bucolic mountain pastures or the cabin in the woods that we long for. We must feed ourselves where we live, from the ground we live on, no matter how damaged that ground is. We must look at ways in which dairying and cheesemaking can be incorporated with urban and suburban life, how we can generate healthy milk-based food in any community. India is ahead of the curve, with livestock being common to urban life. While there are serious concerns around allowing cattle to eat plastic bags out of garbage piles while drinking foul water, I do think we can and must find ways to integrate dairy livestock with dense human settlements. There are ways to heal the false divide between "artificial" city life versus "natural" farm life. We don't need the flood of anonymous milk produced with soulless efficiency in industrial mega dairies any more than we need corporate monocrop agriculture and its toxic crops. We need small- and medium-scale milk production all over the globe, adapted to the conditions of places and to the needs of people, rather than to markets. The answer to our food crisis isn't to continue scaling up. It won't be industrial monocrop megafarms that will feed the future. The answer is gardens, backyard cows, small goat dairies on the edges of towns, networks of dairies, farms, orchards, and pasturelands stacking their functions in and around cities, towns, and villages. The answer is remembering how to feed ourselves and the biotic communities that surround us, wherever we stand on this biofilm that covers the earth in a living mosaic like a mottled rind on a well-aged cheese.

Here, in the most unexpected place, I had found an exquisite cheese that stood for a kind of wholeness and fertility that we can nurture. This was the kind of discovery that kept me going as my mind struggled to connect the intricate webs of terroir that were slowly revealing themselves to me over the last few years of travel and tasting in place. I experienced a gradual and profound awakening in the homes of people who kept the old and novel breeds, who made real food and rebuilt intact communities in defiance of a world that had lost its senses. Slowly, I was regaining my senses. Learning to use them, reclaim them, sharpen and apply them. What I saw, what I tasted and smelled, was the deliciousness that we are all capable of creating when we team with life rather than try to kick it out of the garden. I found the inherent dignity of the human heart contained in a wheel of milk spinning in the moonlight, never separate from its place in the stars.

— CHAPTER FIFTEEN —

Saying Yes to Life

BREGENZERWALD, AUSTRIA

I have seen almost more beauty than I can bear.

—Everett Reuss,
On Desert Trails with Everett Reuss

The trail ran along the narrow crest of a ridge with cliffs plummeting on either side. A panoramic view swept over forests and steep, green pastures, down to tidy collections of houses and barns connected by winding roads to towns in the valleys. The clouds were backlit in somber, purple-red sunset tones as I approached a peak capped with a giant cross. This was one of my last nights in Bregenzerwald, where Austria's western arm pushes into Switzerland and Germany. Small birds chirped *good day* as they flitted past on the mountain breeze. When this breeze blew favorably, the comforting sound of cowbells tolling in the distance floated up from the valley below, in the calm cadence of peaceful grazing. A few remaining snowfields trickled out thin streams of water, like candle wax melting from a flame. I cupped the cold earth juice and drank it down, fortifying my body for the descent through darkness, back to the soft orange lights spilling from the windows of an immaculate white mountain home, where the most profound cheese I have come across was born.

I had tasted the cheeses of Anton Sutterlüty and been struck by their bouquet of explosive, colorful flavors. His cheeses were too good, tragically beautiful. They made me wince in bittersweet ecstasy. Now I was here in Bregenzerwald with Carina Reckers, with whom I had traveled in Sicily, and she had set up our three-week stay with Anton, who agreed to let us work with him in the small village where he makes cheese in the summer. These villages have farmer cooperatives that own creameries built in the bottom story of a home. They hire a cheesemaker for the summer, while they are busy moving their small herds of Brown Swiss cows and cutting hay for the coming winter. Anton is one such cheesemaker, and he ends up buying back the cheese he makes and aging it in Vienna, where he stores some of the wheels of alpine cheese for multiple years—much longer than is common for the regional cheeses of Bregenzerwald called Bergkäse or Alpkäse. His feet walk in two worlds: one in the rural mountain farms, the other in the big city, where his cheeses are renowned by foodies, cheese aficionados, and chefs. He bridges the gap, the imagined dichotomy between the natural, bucolic, and old, and the artificial, chaotic, and new, and his cheeses do the same. They are similar to fontina and raclette, wide wheels with mild smear rinds and a pliable paste, cooked to thermophilic temps in a copper vat. Beyond that, they are edible sculptures that display the possibilities of cheeses as food, art, philosophy, and expressions of personality and place, all intertwined.

Carina and I spent a few nights with Anton and his family in their apartment in Vienna. We ate curry wurst from street stalls and drank impeccable Austrian lagers in halls that served massive wiener schnitzels with cabbage and potatoes on the side. He showed us his aging space, a repurposed cellar hiding under a nondescript house off a quiet, orderly street, the buildings perfectly square and tidy. Beneath brick arches, wooden shelves were overloaded with cheeses from one to five years old resting in the dark. The age range was visible on the rinds, which were maturing from flamboyant orange adolescents into tanned adults with blemished skin, then into brown elders, proud and comfortable in grandfatherly coats of wrinkles, scars, and worn edges. Anton and Carina cut hefty wedges from wheels of various ages, and we tasted through them vertically, from young to old. My brow wrinkled in the joyful grimace that signaled a rising cheese satori. The cheese was painfully, overwhelmingly delicious. It was alive, a collection of passionate voices with too much to say, their hearts bursting as if it were their last day on earth. Whose voices were these? What was this song my mouth could hear?

I listened to Anton speak, a grounded, intelligent man with a gentle voice. His wizard-white hair often sticks out from under a cap, and his face and energy are youthful. He told me there was no climate control in the cellar; its temperature and humidity fluctuated seasonally. He explained that cheeses have historically been aged in similar conditions, not fully removed from annual weather patterns. “It may impart a more interesting, well-rounded character to a cheese, as different enzymes and ripening processes kick in at various temperatures. It's good for a cheese to have some inconsistency." I'd never heard a cheesemaker say this. Rather than attempting to control the aging processes by maintaining an artificial climate of consistent temperature and humidity, Anton lets the cheese roll with seasonal cycles, interacting with the world outside the cellar. He's not completely taking his hands off the wheel, though. If the cellar feels dry, he pours water on the floor. The weekly washing of the cheeses keeps the rinds from drying out and introduces humidity into the room. Most important, he keeps the room full of cheese and wood shelves. Wood absorbs or releases moisture to help buffer fluctuations of humidity. The air and wood carry the microbial critical mass of the cheeses and their rind community. This community is formed and maintained by washing with brine to encourage a smear to form. He doesn't measure the salt, but it's probably around 1.5 percent, which forms a mild smear that doesn't break down the paste or lead to bold flavors. It's the same salt percentage that the cheese has, so no additional salt will migrate into these wheels that may be washed three times a week for five or more years.

Anton's process for selecting which wheels to age long-term is intuitive and hard to put into words. Some clues appear early on, such as how the wheels look coming out of the brine and what they do in their first few weeks, but he admits that it's hard to predict where the cheeses will go as they age. A cheese might be underwhelming at three months, delicious at one year, and then not advance further, or even go downhill from there. Even the four wheels from one batch can turn out different. What impressed me about this first of many heady conversations was how humble he was about the process, how little credit he took for the cheeses, and how he embraced and even celebrated the mystery, the lack of predictability or control. He seemed less interested in making cheese with consistent results, and more in the art and beauty that emerge serendipitously from the myriad complex pathways available to cheese when we allow it space to grow as it will. Anton's cheeses speak for themselves and reflect his message that "we can't improve upon the designs of nature. All we can do is let the natural world flow through the creations of our hands and bodies."

Before we headed up to start the summer cheese season, we needed to pick up rennet. A small-town butcher invited us into his workspace and brought out large, dried abomasa hanging from meat hooks. Wearing a bloody apron, he cut them off and threw them in a box. They felt like crinkly leather and smelled of fish sauce and organs. Three- to four-month-old calves were brought in by local dairy farmers to become veal. The abomasa were inflated like ballons and hung unsalted in a warm, dry room for two weeks. The approach to rennet favored in the Alps and Northern Europe is to use just the skin of the organ to create a clean liquid rennet with little lipase. This contrasts with the rustic whole-stomach rennets more popular in the Mediterranean, where the internal cheese is used along with the skin of the organ to make a paste that leads to the bold, spicy lipase flavors that characterize many Italian cheeses.

We stopped at a little hardware store that sold cheese supplies and equipment, and I was perplexed when Anton purchased a jar of 100 percent chymosin FPC rennet powder. He explained that sometimes the rennet he made wasn't strong enough and needed a bit of a boost. The dried abomasa were cut into strips and mixed together, then the amount needed for one week was covered in whey and steeped overnight at 105–120°F (40–50°C). The whey became quite sour and extracted the coagulating enzymes from the strips, which were strained out of the liquid rennet. Anton tested a new batch of rennet by dropping a few milliliters in a measured amount of warm milk and counting how long it took to coagulate. If it took longer than eight seconds, he added some FPC powder until it reached the proper strength.

All geared up, we headed up the mountain past ski resorts and fancy chalets, through tiny villages that each had their own cooperative creamery and were surrounded by unfenced green pastures.

Gebsenkäse

Every evening, the farmers of the cooperative brought their milk to the creamery that filled the bottom story of a house, partially buried in a hillside. The milk was weighed and poured into shallow, wooden five-gallon tubs called *gebsens*, where it sat overnight. The gebsens were placed in a row on a bench, and multiple offset layers were stacked on top of one another in a pyramid. The milk was left overnight in the cool basement creamery, and skimmed in the morning, with minimal souring taking place. For Anton, the gebsen and other tools are an essential part of the cheese, and the farmers who see him still using these are pleased and proud that this aspect of their heritage has not been lost or relegated to a museum. He feels the tool is so integral to the process that he refers to his cheese as *gebsenkäse*.

When we came down the stairs before sunrise each morning, my first task was to run a finger around the edge of each gebsen, loosening the cream. Sucking the cream off my finger, I looked for a clean, sweet taste, with just the slightest awakening of malty lactic flavors. The cream was removed quickly with a perforated metal scoop that allowed the skim milk to fall through. Each gebsen was lifted and poured into one of two copper vats. A wheelbarrow was filled with wood and left outside the door as a fire was lit under one vat. The farmers arrived just after sunrise with udder-warm morning milk, which was poured directly into the vats, blending with the skimmed evening milk.

The milk was cultured with whey from the day before, which was heated to 145°F (63°C) and then held at 100°F (38°C) before being refrigerated as an active, thermophilic starter. Like other natural starters (clabber, kefir), it must not overferment or it will run out of food (sugars) and become inactive. This whey was added to the warm milk in the vats, one milliliter for each liter of milk or 0.1 percent, a microdose. For comparison, I use 1 to 2 percent when I make cheese with any natural starter. After the day's cheese was done, we collected whey to become the next day's starter, carrying forward the milk-fermenting microbes from one batch to the next. Rennet was stirred in just before we went upstairs to have a hearty breakfast.

The curd had barely coagulated when Anton began to cut it using a tall harp. He held the implement like a dance partner with whom he followed ritualized movements that turned the delicate gel of the coagulum into rice-sized pieces. A mechanical stirrer was dropped in and the fire stoked to cause a rapid heating of the copper, bringing the milk quickly up to 125°F (52°C). Anton picked up handfuls of the curds and ate a few as he squeezed them into a ball in his hand, observing the whey as it trickled off his wrist. Stirring continued until the curds squeaked on the teeth and emitted clear whey. Too much moisture in the curds and the cheese would ripen fast and be unsuited for long aging; too little and it would result in a drier cheese that also wouldn't ripen properly. The narrow middle path was found by Anton's hands, eyes, and mouth, tuned by years of repetition, lending an intuition that could never be transferred into a recipe or fully explained in a book or video. His whole body was calibrated to the process, and his subjective tests produced feelings and sensations rather than objective metrics like moisture content and pH. Such numbers are useless without a body and mind focused on crafting deliciousness, without a heart that beats and is in love with existence.

Stirring stopped, and the curd settled as a conical mountain at the bottom of the vat, hidden from our sight by the sea of whey. Anton wrapped one edge of a thick hemp cheesecloth around a flexible metal band held as an arc between his arms. The two opposing corners of the cloth were clenched in his teeth. Bending over the vat, he moved his arms into the hot whey; the cloth ballooned like a fishing net as it was pulled through the curd mountain and brought to the surface. The four corners of the cloth were tied together and a hook on a pulley was used to hoist the heavy ball of curd out. This pulley slid on an overhead track, transferring the hot, dripping mass to a draining table, where it was plopped into an adjustable-diameter plastic ring resting on a wooden board. The process was repeated until four curd balls rested inside rings. A second board was placed on top, sandwiching the balls as they slowly morphed into wheels. Pressure was applied by a screw jack mounted to the ceiling, and the wheels were flipped periodically throughout the afternoon.

The whole room turned into a steam bath, and we sweated in our rubber boots as Anton closed the windows and cranked the fire up to heat the whey for ricotta. The whey was very sweet, so once it reached 185°F (85°C), some acid had to be added to cause ricotta to form. Large wooden barrels full of sour scotta were kept in one corner. Anton added a few scoops to the vat full of whey until a firm ricotta formed. There is no market for this ricotta, so it was fed to the hogs on a neighboring farm. What Anton was after was the remaining clear, greenish whey, the scotta. All the gebsens we used were spun in this hot liquid while being scrubbed with a brush. It was also used to replenish the barrels of sour scotta, to feed more sugars to the community inhabiting that medium. Besides the citric acid used on the copper vat, this sour green whey was Anton's only cleaning product. We used it to wash the cheesecloths, the floor, and the wood boards the cheese was pressed on. After the first batch of cheese of that season was made, we used it to wash the wood windowsills and tile walls.

Some might consider Anton's process to be sloppy, but it was actually a highly choreographed and responsive dance with microbes, the cultivation of a particular type of bacteria as the basis of a time-tested food safety program. The scotta had been heat-treated and then cultured in the barrels by a community of thermophilic LAB similar to what existed in the whey starter. By bombarding the room with the sour scotta and washing all the implements in piping-hot whey, he was cultivating a culture in the whole creamery and maintaining a balance conducive to his cheesemaking. This was not a free-for-all; the process was altered to influence the fermentation of the cheese. When weather got hot and the cheese fermented too quickly, he washed the gebsens more thoroughly and scrubbed down the pressing boards with steaming scotta more frequently. He said the first few batches of each season usually don't turn out, probably because the microbiota of the creamery has not established yet. This is why he had me wash the walls and sills with sour whey after our first batch. The room itself was a carrier of starter culture, and an amplifier of terroir, and we had to get that up and running.

Rather than attempting to prevent all microbes from growing in the space and on the equipment, he encouraged a thriving population of the bacteria that fermented the cheese, feeding them daily through the act of making cheese and ricotta so he could obtain more whey to maintain a healthy seed bank in the soil that was the room and its tools. When the farmers walked through the creamery in muddy boots, any problematic microbes wouldn't have ground to grow in because the field was thick with the community of beneficials who defended their turf. It stood the logic of the industrial antibiotic paradigm on its head and produced one of the most tantalizing and expressive cheeses I have come across, by teaming with microbes rather than attempting to harness, purify, and police them.

Anton was facilitating an environment that housed positive resident microbes rather than obsessively bombarding the space with chemicals in a neurotic and futile attempt to remove life. Rather than trying to limit the variables, he was actively increasing them, giving more room for the river of life to flow into the cheese. Rather than using a single package of regimented bacterial strains, he introduced to the cheese a multitude of microbes entering from various directions: the raw milk sitting overnight in wood, the whey starter, the rennet, the wood boards, the sour whey used for cleaning, his arms dipped deep into the vat to pull the cheese into the world. Rather than sealing himself in and attempting to sterilize everything that came into the creamery, Anton opened the doors and windows. The cheeses that were shaped in this space by Anton's rigorously tuned body were beyond safe and good. They represented a whole other way of living, of being, of belonging to and honoring the webs of life rather than fearing and suppressing them. As Anton said, "The question is, are you going to say yes or no to life?"

Letting the Milk Sing

The make process was a flurry of frenetic busyness. Anton's calm and meek demeanor transformed into a whirling blur of multitasking. If my memory is correct, he grew four extra arms. He became another person: serious, speaking loudly, giving firm orders. As I struggled to match his motions and stamina, he corrected me in a direct manner. The motions involved physical exertion, flipping heavy wheels, stirring aggressively. A massive amount of kinetic energy was transferred into the cheese; he was pouring himself into it on many levels. He knew exactly what each motion should look and feel like and was conducting a fast-paced orchestra of the daily events in the creamery, where timing is of the highest importance. It was remarkable to see these two sides of him, and I respected him highly for taking his craft so seriously. People who make transcendental cheeses share a drive that is passionate and obsessive. I have met a few makers whose cheeses channel aspects of themselves; their presence is embedded in their work. Their cheeses taste like their personality, philosophy, or energetic essence, even their madness.

Anton practiced a singing style that involved letting a song bubble up and emerge spontaneously from his body. He would start humming a tune, voicing whatever sounds came naturally to him at that moment. It was an externalization of internal states, giving voice to various aspects of his being—emotions, sensations, the condition of his spirit and psyche. Sometimes the song evolved into something beautiful, full of joy and love. Other times it was sad, melancholy, angry, or anxious. Sometimes not much happened at all; the song never grew into a meaningful expression. He would just let the song come and go, without forcing, judging, or resisting its momentum. Songs are transitory moments, expressions of what it means to be alive for a few precious moments as a breathing, feeling primate with spiritual leanings, graced with the miracle of walking on an enchanted planet full of sorrowful love and frightful beauty. We can all sing like Anton. These songs sing themselves, with our bodies giving them shape and birthing them into the air before they fade back into the radiant silence.

This extemporaneous singing is an audible manifestation of Anton's philosophy on life, his worldview. Our emotions, moods, and spiritual health vacillate, shift, and meander like a river through the floodplains of our days and nights. We seek to maximize happiness, to prolong the states that feel good; and we avoid those that feel bad, scorning the sadness or emotions we label as negative. When we resist the flow of life, we trip and lose the power of the moment. Like the songs, our moods change, rise up and move through us as temporary waves of expression, and move on. "Life is moving, let yourself move with it," Anton gently suggests. "Allow yourself and your art to be carried away by life's currents."

This therapeutic singing, this philosophy of acceptance is embodied in the cheese. Anton's art is to "allow the milk to sing," to express the unique moment in time when the cheese emerged from a place, through the body of an animal, in a vat, in a room washed in whey. He starts with raw milk that contains a multitude of voices, a choir of microbes. The voices of the fields surrounding the house, whose blades and blossoms are transformed by the bacteria residing inside cows. The voices of the families who care for the cows, the hands that milk them, the sheltering walls of barns. This choir is backed by an orchestra of multiple sections: the rennet made from calves, the sour whey in the barrels, the wood pressing boards, the gebsens with their rough surfaces and mineral deposits. The firewood cut from the surrounding mountains, the spring water flowing from the tap. The house itself becomes an amphitheater in which the song builds strength. These cheeses are operas written by life, conveyed through the hands of a humble maestro who is coordinating the sounds of a thousand musicians into a cohesive tribute to the present moment. Not forcing but allowing the song to emerge from the orchestra. An entire drama can be told, sung, birthed in a wheel of milk.

Anton cannot say what a nascent cheese will become. Like his songs, his cheeses are spontaneous creations. Sometimes what emerges is a poignant expression of how wonderful life can be. At other times it takes strange or less desirable turns. As Anton puts it, "You never fully grasp how sometimes it just happens, and other times it doesn't."

Speaking with Microbes

Anton spoke of his relationship with microbes in touching and insightful language that moved the mountain of my soul. "The microbes have to get to know you," he said when discussing the process of learning to make cheese, which for him involved communicating with microbes, learning their language, establishing rapport. He rarely mentioned specific types of microbes but spoke of a collective entity whose language could be understood as a way to open a dialogue with life, the universe, or God, whichever word we choose to use. We should approach this entity humbly, slowly learning to discern its subtle presence in how the milk transforms, attentively observing what paths it chooses as it matures into a cheese with infinite possible futures. According to Anton, this observation is not a one-way street: "The things we look into intently look back at us. As we work intimately with microbes, they are working on us." Microbial consciousness is not passive or inert. As we gaze into this other dimension, mesmerized by the milk swirling over the crackling fire, something gazes back. As we reach into the vat with open hearts and hands, something reaches back.

Milk microbiota don't use tongues to speak. We hear their whispers with our hands as we squeeze the hot curd and feel their invisible elven script on the bottoms of the wood pressing boards. We can smell the residue of their poetry on the cheesecloths, see evidence of their life cycles as we remove cheeses from the brine. Through the repetition of everyday rituals and chores, the artist cheesemaker builds a sensitivity to the fermentation that is the song of the microbes. You start to feel the milk shift seasonally, with changes in weather, who is milking, which field the cows are grazing. You can observe slight changes in the color of whey, the feel of the curd, how the cheese responds to pressing. The next step is to learn how these things lead to different characteristics in the cheese. You take notes with each batch and watch how the cheeses age. After years of doing this, you start predicting changes in the milk and cheese and preemptively shifting things in the make to compensate for foreseen imbalances. This could mean using more or less rennet or culture, leaving the windows open or closing them at night, changing your cleaning procedures to soothe the microbes and ease the daily process of birthing cheese.

As I walked this trail, I slowly recovered from the training that had taught me to view microbial continuity as a source of trouble, as something to be broken. I had been taught that we need to erect barriers between the farm and the creamery, between the creamery and the aging space, between our bodies and the cheese. To make consistently good cheese, we needed to reduce and control the variables, to keep the dirty world out, and to create a fluorescent-lit sanctuary inside a white-walled fortress where we can play God, locking the gates and enforcing a strict guest list of who is allowed inside. From the walls of our fortress, we keep the microbial wilds at bay, kill all the wolves, pull the weeds, set poison traps, cleanse the castle of spirits. We and the cheese we make are separate from the messy chaos of existence, which always lurks, pacing just outside the walls, searching for a way in. We will impose order, straighten every river, and level all the mountains, so we can finally be safe and free, in our cold, lifeless fortress. As paranoid insomniacs, we always sense an imminent invasion, the howl of wolves keeping us restless at night. My years of unlearning and healing from this cultural delusion culminated with Anton laying out his philosophy and inviting me to make cheese with him, to experience the power and eloquence of microbial creation.

Anton said that he was working to invite as much of the outside world into his process as possible, embracing the constantly changing cycles of nature in a dance. If you allow the pulsating currents and transitory moods of the weather, the cows, the land, and the year into the creamery, cheeses can be born that are manifestations of flourishing life. Letting the outside in, allowing a continuity to exist among the livestock, farms, farmers, fields, barns, milk cans, wooden gebsens, aging boards, barrels of sour whey, shelves where the cheeses age, and hands that wash them was a route to great cheese. There is no kill step, no line drawn in the sand with us over here and the chaos of unknown microbial wilderness over there. Life is allowed to flow from the farms and fields through the milking parlors into the creamery, into the cheese, into our bodies. Not carelessly, but with observant diligence and intimate participation.

He talked to the farmers about how they milked, tracking where the cows were grazing, how the health of the herds, pastures, and families were. When it rained and got muddy, the milk changed; everyone had to be more careful in how they milked. Rather than pasteurizing the milk or keeping the farmers and their muddy boots out of the creamery, he worked with these cycles, adapted his process, but allowed the land and herd to continue voicing their mood. He dried his cheesecloths out in the sun while bringing in a load of firewood from an old shed that had never been tested for pathogens. When we open the door and welcome life into our lives, into cheeses birthed by houses of biological richness, what emerges is a cheese that sings a beautiful, woeful lament that also contains a message of hope, of the sunshine just behind the clouds. The seeds of a better future sit in the soil, ready to sprout after this long winter, after our fortress of sterility and hubris crumbles back into the earth. It is time for us to reconcile with the world, to leave offerings to feed our neglected mother, who is so thirsty for our love and hungry for the gifts that only we can craft.

— CONCLUSION —

Reimagining the Future of Cheese

CASCABEL, ARIZONA

My favorite music is the music I haven't yet heard.

—John Cage

Returning to North America, the continent where I was born, all the experiences, lessons, and bodily memories of my travels aged in the cellar of my soul. I continued cheese trekking by spreading the seeds I had been gifted, as humans across America and abroad—hungry for regeneration through authentic connection to land, microbes, and health—invited me to visit and teach in their communities. Wrapping up a camper-van circuit of the western states, during which I offered five-day workshops in collaboration with host farms, I landed in a tiny community called Cascabel, far down a dirt road in the Sonoran Desert near Tucson. Spending the winter there cloistered in a trailer, feeding horses and milking goats, walking at dawn among ancient saguaro cactus, following deer tracks along the dry arroyos, and visiting sacred spring-fed creeks, I sat down and devoted my days to putting down these words. It has felt amazing to release these stories and experiences from my body, leaving me ready to move on, my heart and backpack a bit lighter now. The future is chaotic and uncertain, but I trust the journey.

As much as I love to learn about well-established but never static cultural practices of milk fermentation, what really concerns me are the cheeses of the future. My work is less about preserving tradition and more about innovation to realign our current approaches with what has been successful in the past. We can collectively reimagine what sane food systems look like, as we relearn how to be an integral part of biology on the rind of a living, breathing, sentient planet. Simply copying the cultures and cheeses of other places and times is not enough. We can craft new traditional cheeses, in new places with different milks, and build a terroir that conveys our time, what our culture is going through now.

Years of cheese trekking have pointed me toward solutions to the problems I have identified. These come not in the form of cheese recipes but as a composite approach to rebuilding microbial continuity in our milk foods, and a vision of how terroir can be built again, established in places it has never existed, in cheeses the world has not yet tasted. Unlike the terroir of the past, which grew slowly over generations of trial and error, building the rituals that worked and then sticking to them, the terroir of the future may arise in other ways. It could be based on microbial analysis, testing, and published data about how farming practices, equipment, and the conditions of cheesemaking and aging spaces impact microbial communities. Natural cheesemaking can be shown to be not only safe, but safer than the antibiotic approach, and this validation can help shift the paradigm and strengthen our knowledge of the seeds and how to cultivate them. We have access to the technical know-how of traditional cheesemakers from around the world, and can reinterpret those techniques wherever we are while acknowledging the people and places who inspire us. I do think we should choose styles that fit our farming systems, climates, and economic necessities, but with an eye toward designing foods to feed the rapidly shifting future. We no longer need to be, and in many cases can't afford to be, bound to tradition. The lack or loss of traditions can be used as an advantage, as a space in which we can innovate.

We can build bridges between the rational, materialist, scientific worldview and the mystical understanding of the sacredness of life, the intuitive touch of the artisan, the energetic communication among us, microbes, livestock, and landscapes that can never be empirically measured. Both perspectives have validity, and they are not mutually exclusive. Science and the use of our brains have always required a heart that beats and a body that speaks. We need to find a balance, a science that acknowledges its limits, that is nestled within an understanding of the sacredness of life, that rejects scientism. We can learn to listen to microbial communities while analyzing the composition of species in them. We can start making the wood tools and using the copper vats and woven baskets, and improve our art by knowing how these materials impact the cheese, how to use them to tweak our process.

This isn't a time to pick sides: either science, knowledge, and measurement or spirituality, intuition, and art. This is a time to transcend that limited, dualistic perspective, embrace uncertainty, and realize that most things exist on spectrums, and some things lie outside the light shone by science and yet are very real. We need cheesemakers, farmers, companies, educators, and consultants to exist all across the spectrum, to allow for experimentation and the blending of approaches. Some can stick completely to tradition, keeping the old breeds and tools in use. Others can innovate, adopt, and apply modern scientific perspectives. Makers operating outside commercial production can practice natural cheesemaking as a demonstration, and work in tandem with universities to test its results empirically. Some can teach how to cultivate and commune with microbes without resorting to Latin names and the limited perspective that sees them as fixed and final. There is room for all of these approaches. There is no one right way to make cheese, to farm, or to live.

The beauty of cheese—and to me, its most fascinating quality—is its diversity, the infinite combinations of texture and flavor that are possible and how these can express a diversity of landscapes and human cultures. This diversity is quickly being eroded, but we still have space to record what remains in isolated pockets, and turn the corner from eradicating to revitalizing and reinventing sane foodways. The established seeds of livestock, microbes, and approaches to cultivating and partnering with them are in our hands. In small, forward-looking cultural islands, these seeds are being planted in new gardens, soils built up to support them, microbial alliances resprouting as we remember how to cultivate rather than control life. How to stop plowing the fields and applying chemicals and instead nurture life, allowing the mycelium to spread under the soil, through our built spaces, into the creameries and aging rooms, onto cheeses, and into our bodies, which long for the lost connection.

Throughout the preceding pages, I've been beating around the seeds of cheese. In talking a circle around the thing, its form and contours have emerged, and now we can see its full beauty and speak about it directly. We can now look explicitly at the seeds on the table, sort them into overlapping categories. We can explore how each one can be renewed, revitalized, and planted again as we adapt the seeds to our gardens of the future, which will feed us and the planet from which we sprout, to which we will be composted, becoming food for life again.

Microbial Cultures

I'm not so interested in taking the microbial cultures developed in other places and bringing them into our cheese. That would be literal cultural appropriation and bioprospecting of the microbial diversity maintained by the remnants of pastoral societies around the world. Heirloom cultures are transitory; they shift and should be allowed to do so. The fact that they are undefined and unstable is a gift that should be celebrated, for they can adapt to changing conditions, and their diversity makes them resilient. Extracting specific microbes that we think can serve us and attempting to freeze them in an unchangeable state is the basis of the corporately produced cultures of microbial imperialism. Rather, I advocate for cultivating the microbes indigenous to milk, working with the microbes found in the milk where we live, farm, and make and age cheese. The process of farming and fermenting milk selects the microbes and makes the cheese what it is. Cultures can be cultivated specifically to a place and its people, and for the cheeses they co-create.

There is the issue of missing microbes due to the steady and intentional degradation of microbial biodiversity in milk and cheese. To address this, we should begin by looking at how we can change our farming, milk storage, and cheesemaking practices. We can rebuild the lost diversity and establish terroir by looking to the less obvious seeds on the table. Rather than a quick fix of mining the microbial commons and splicing that genetic wealth into the irredeemable disaster of industrial agriculture, we need dramatic structural changes in how we farm and feed our communities.

The first practical step is to find out which microbes are in milk and how we can encourage or discourage them to grow. I suggest a citizen science project similar to what has been done with sourdough. People around the world can clabber their milk and have it analyzed, so we can see if there are geographic patterns to microbial diversity, and how various practices select for certain microbes. Kefir grains could be collected from various places and similar research done. A kefir library could be established to give us a sense of the diversity of microbes in these communities, and how we as keepers of these seeds can best care for them, work with them to ferment milk and build cheeses. The ecological succession of microbes within individual cheeses has been studied many times; what is missing is a broader, global look comparing this data.

The second step is to overturn the concept that having low levels of bacteria in raw milk is inherently good. This requires a shift in how we consume milk. When the aim is to keep milk sweet for as long as possible, bacteria become a problem that we invest enormous amounts of energy in fighting. We need to embrace the souring of milk, allowing healthy populations of bacteria to carry milk toward its destiny. This means consuming sour-milk beverages such as kefir and clabber, and looking at milk as something that is cheese waiting to happen. Incentives could be offered for the production of cheese milk, and for farming systems that lead to healthy, microbially rich raw milk.

Livestock

Livestock serve as the bridge, or the conduit, between landscape and milk, and the link shifts in response to how we farm. Livestock connect us to the land, and the land to the cheese. This moves us from microbial to animal terroir: how the breeding and raising of animals shapes and influences a cheese.

Breed Genetics

The genetic diversity embodied in the heritage livestock breeds of the world represents an immense wealth that is being decimated. To my mind, the loss of this wealth is more tragic than the loss of the cheeses. Some of the lost cheeses can be re-created, and new ones developed, but once the old breeds are lost, they are gone forever. Generally, the loss is a slow process of genetic dilution, making it hard to perceive. Efforts to protect and revitalize these breeds are important and should be pursued, especially where the efforts are guided by the communities who have historically kept these animals, with a deep knowledge of husbandry and breeding that leads to breeds with a healthy level of genetic diversity. While I applaud some efforts, others go too far, reflecting notions of racial superiority and eugenics. A gene pool that is too closed and shallow results in genetic bottlenecks and the loss of what makes these breeds resilient. Wise and knowledgeable keepers fine-tune the balancing act of maintaining a breed identity while allowing enough genetic exchange for it to remain resilient. Heritage breeds that are still identified with a place and their cheeses are living manifestations of the symbiosis that we are capable of, that the cheeses in this book represent.

In contrast with my feelings about the bioprospecting of microbial cultures, I'm not opposed to taking breeds from one place and introducing them somewhere else where they are likely to thrive, or blending breeds to create something that suits where we farm and our goals. Borrowing a concept from permaculture, I think we should move away from the false dichotomy with racist undertones that there are native and introduced species, and start breeding for the future, for what climates, ecosystems, and economies could shift into. Rather than only using the breeds from a region and attempting to purify them to an ideal state, we can bring in new genetics via traditional crossbreeding methods, and begin building diverse breeds to fit our predicaments, just as humans have done for countless centuries.

Storing genetics in banks is a necessary step, but not enough. Like the freezing and making static of microbial communities, it represents a misguided approach of control and stability. The breeds must be bred, not just for the sake of maintaining them, but to improve their lives as working animals on working farms, where their genetics can radiate out with different lines growing and shifting to meet changing conditions and needs. The rare breeds can't be kept in a zoo. We must raise them, eat them, milk them, allow them to live and do what life does. Change, adapt, proliferate, diversify.

Our aim should be to breed and farm for cheese milk rather than for the overabundance of nutritionally poor liquid milk. Consuming large quantities of sweet liquid milk, raw or pasteurized, is a misguided luxury of the age of refrigeration, with puritanical overtones of the fear of fermentation, of mold, of life and death. We can breed for traits that were valued historically for obvious reasons: resistance to disease, ability to thrive on low inputs and navigate rough terrain, resilience to local climates, less mastitis. Rather than focus on volume, we could breed for lower yields of richer milk, factoring in the true cost of the inputs, such as the feed and medicine needed to maintain a healthy herd. Animals may produce less milk but it would contain higher levels of nutrients and flavor, and this would make more cheese with higher concentrations of deliciousness. We can aim for milk that contains less water and more fat, protein, vitamins, and minerals, and has higher concentrations of flavor compounds expressing the landscapes and the microbes of its origin. In short, we can breed with the aim of generating terroir, of crafting unique place-based flavor in nutritionally dense foods. Milk can be a uniform, faceless commodity, or it can be a nuanced, richly satisfying food. Milk is not just a single thing. There are many milks.

Teat Ecology and Treatments

Livestock are the source of most of what ends up constituting the milk microbiota. We can adapt our farming practices to aim for increasing microbial diversity in milk, especially in regard to the microbial ecosystem found on teats. Rather than treating this ecology as a field full of weeds to be sprayed with chemicals, we can allow it to be a healthy home to the microbes beneficial to cheesemaking, which is what it naturally gravitates toward in healthy farming systems. Rather than the antibiotic products commonly used on teats, which shift the balance in favor of problematic microbes, we could either not use any products or choose probiotic ones. The natural starters used for cheesemaking (whey, yogurt, kefir, clabber) can be used as teat dips, carrying the culture backward and feeding it at one of its sources. There are many non-iodine teat dips that utilize lactic acid, such as one commonly used by cheesemakers in Norway, a low pH teat product designed to encourage positive LABs. Another approach is the use of wood wool, coarse shavings that can be used in place of towels or rags to wipe debris off teats before milking. Wood wool has been shown to have positive impacts on teat ecology and can be composted along with manure after use. These approaches to teat treatment flip the script and let us begin allowing the seeds from the landscape and barns to transfer into our milk and cheese once again. Rewilding the milk microbiota could start with this equivalent of removing dams and allowing salmon to return upstream as efforts to restoring ecosystems.

There is no way to do this within the industrial system of huge herds, the pooling of milk, and long-distance shipping. That system is inherently unsafe and unsustainable. We need small-scale milk production everywhere, which could be subsidized. It's already happening on the fringes, and those who provide healthy milk and cheese for their communities should be celebrated and supported, rather than hounded and villainized. There is no saving the industrial corporate food system. We need to start taking it apart and cleaning up the mess it has made.

Landscapes

Botanical Terroir

Zooming out from microbes and livestock, we now can point to how wider landscapes and ecosystems are sources of microbial and botanical terroir. When animals are raised on pastures or rangelands, the diversity of plants they eat and sleep on can translate into flavors in cheese. This is usually not a direct transfer of plant characteristics unless it is a strong aroma like garlic or mustard. It's more common that a diverse diet of living plants leads to a hard-to-describe depth in a cheese, or that plant terpenes impact the microbiology of the milk. When hayfields are cut and dried, the plant components are transformed in another way and provide the basis for winter or dry-season milk.

The way livestock feed is grown can be altered. We can move toward local, organic production of hay and grains, utilization of wild trees, orchards, and fodder crops such as fava, turnips, or sorghum. Hayfields are often monocrops, so their diversity can be restored, and we can resume the practice of cutting and storing dry hay. Most important, we can look at ways to take the animals to the food where it grows semi-wild as forests, rangelands, and deserts. Extensive grazing can be accomplished in almost any ecosystem in a manner that is regenerative. We can get animals back on the land and reintegrate our human lives in biological webs as we embark on a new pastoralism based on the wisdom of the old. Many human hands will be required to produce food this way, and the people who are interested in pursuing a pastoral life should be encouraged to explore it, and paths developed to make this lifestyle financially feasible for them.

Barnyard Terroir

Potentially more important than this dietary aspect of botanical terroir is the impact of the bedding material that livestock sleep on in barns, especially during seasons when animals are kept mainly inside. The material used is most commonly straw remaining from grain harvesting, sawdust, or wood chips, but I have seen pine needles, dried leaves, and branches used as well. Each of these materials carries its own microbiota, which is modified as manure and urine are introduced and composting begins. This complex bedding microbiota interacts with the teat ecology and connects the landscape to the cheese in a direct way. How humans manage this bedding and other conditions in the barns and living spaces shapes the health of raw milk.

If animals are lying in mud or their own manure, this can create problematic milk. If bedding is changed out regularly, or more material is added to build up what is called deep-bed compost, which generates heat, this can have positive impacts on the milk microbiota. When the bedding is obtained locally, a continuity is allowed between surrounding fields and the spaces where animals live. The plant material, dung, and urine are then composted and ideally returned to the land as fertility, which can also contain the literal seeds of the plants the animals consume. A circular flow is established, the barnyard engine of terroir, which combines the inputs of microbes, livestock, plants, farms, and farmers.

Water

Water quality and composition are an understudied and overlooked facet of terroir that should be considered a seed of cheese. Milk is mainly water, so the source of this water, the materials used to contain it, and its constituents necessarily play a role in milk microbiology and therefore cheese. Seeing livestock drink spring water from stone troughs or cast-iron bathtubs, wood trough, or streams has caused me to consider how the use of natural materials could be a consideration expanded outside the creamery. How do the minerals in the water impact animal health and milk properties? This problem has mainly been looked at in its negative manifestations, how water can be a source of pathogens or contaminants, but I've seen little research into beneficial aspects. The quality of the water exiting the farm is another indicator of the overall health of this system. For dairy farming and pastoralism to be ecologically sound, we must consider its impact on waterways and sources, and realize it is not inherently negative.

Also relevant is the quality of the water coming out of the tap, which is used to wash equipment and rinse floors, and can also be intentionally added into the cheesemaking process in the form of rennet or styles that involve curd washing, as well as rind treatment. Not surprisingly, the modern approach is to use reverse osmosis (RO) or UV light treatment, the equivalent of pasteurizing the water. Which is not always a bad idea, but nor is it always necessary. To reduce water to simply H_2O, and remove even its minerals in the case of RO, is in line with the industrial approach. Water has been and still is used without these procedures in many places, where people trust their water, or boil it, or are very careful not to allow it into the cheese. Is water another seed of cheese, a vector of terroir? This is a question worth exploring.

Human Cultures

Cheeses develop microbial terroir because certain members of the milk microbiota have been selected for through the variables of farming, milking, and making cheese. There can be strains particular to a place or farming system, but these are generally individual expressions of broader types that can be found everywhere that healthy milk microbiota are allowed to flourish. What really turns this microbial ecosystem into a particular cheese is how the humans involved influence it. This cultural level of terroir is where it all comes together. We need to acknowledge that it is not a recipe or machine that makes the cheese. It is us. We are facilitating a fermentation and transformation, and this role should be taken seriously.

One aspect of this is the tools we use, how they are made, and how they shape the final cheese. Tools that have been used for decades take on a patina of significance, imbued with meaning and spiritual value. When tools are handed down from one generation to the next, or from master to apprentice, they shine even brighter, with a light that a generic, mass-produced piece ordered from a website could never match. When the tools used are crafted by the cheesemaker, or by craftsmen down the road, they contribute their own terroir. Tools made of local materials can carry an additional sense of place into a food, like a special background spice in a cake that you may not actually perceive. But it's there, making the cake more deeply satisfying and expressive than it would otherwise be.

These tools can once again be made from natural materials, supporting the artisans who produce ceramic or woven forms, wooden vats and stirring implements, metal or wood implements, copper pots, and wood aging shelves. Rather than choosing from a limited range of standard plastic forms to shape our cheeses, we can give them unique character by having custom forms made from materials that can be homes for microbes, helping establish rind communities and microbial continuity. A wood vat can be the reservoir of the starter culture, as it absorbs whey and releases into the next batch.

The buildings of a farm are another aspect of cultural terroir and can be conduits of microbial continuity. Like the tools, we can construct these spaces from locally appropriate materials and allow microbiota to exist on their surfaces, which interact with those of the surrounding landscape, the livestock, and our own bodies. How barns and milking parlors are cleaned is a part of cheesemaking. Following rigorous protocols, paying close attention to the animals and the milk with an awareness of how the process influences the cheese, also allows potential issues to be nipped in the bud, or for precautions to be taken in the creamery to address the observations of those who milk. Rather than viewing the creamery as a space that needs to be kept free of microbial life, we could treat it as a workshop with conditions conducive to their growth, an indoor garden where microbes are cultivated. Less hospital, more kitchen. The regulations of how these spaces are constructed and the chemicals used to clean them can be relaxed if we address the health of milk upstream, in the fields and barns. Wood and other porous materials that can harbor microbes need not be feared; they can in fact be part of a food safety program. Implementing these suggestions may seem far-fetched to some, but truly the only thing standing in the way are governing bodies and their legal requirements, which are far from fixed, and unfortunately not necessarily based on sound science.

Lastly, we can view the practice of cheesemaking as being experientially learned and encourage those entering the craft to apprentice with makers whose approaches and cheeses appeal to them. Rather than focusing on studying dairy chemistry or assuming we can develop and follow recipes to achieve certain cheeses, we can teach the craft through a more holistic lens that combines science with sensory observation and thinking in terms of the wider farming system from which the milk originates. Since most of us don't possess the bonds that unite human communities to cheeses, we will have to rebuild them. I suggest a cultural exchange, wherein people who are interested in dairy farming, pastoralism, and cheesemaking are given opportunities to visit and learn from traditional and innovative farmers and makers in other parts of the world. Those who are tied to farming and making cheese in one place can open doors to apprentices and knowledge holders from other countries. There is room for cheese consultants to be a part of the cultural shift, as well as cheesemongers and anyone involved in the world of cheese. Education, knowledge, and training—as part of living human cultures—make up the hidden foundation of terroir in cheese.

— POSTSCRIPT —

The Passing of the Seeds

Milk and cheese have been my path to healing, to finding my medicine. By allowing something I was curious about to consume my life, I was flung out of apathy on a journey that shook something loose in my soul. As I walked, the courage to speak grew and, in speaking, I found my voice. Following the impulse to take that job in Mongolia, which my rational mind knew was a bad idea, set me on the course that has culminated in this book, which feels as though it's been writing itself. This book was not born from a logical plan or a linear outline conceived in my brain. These words arose from somewhere that can't be mapped or seen, from some power deep in the ground that was pulled up through my bare feet and spine into my head, then trickled out my fingers.

Cheese was the lens I picked up and looked through to see the world and some of the myriad ways we can inhabit it. Following the golden thread that I found in a glass of milk, I was reconnected to the cycles of the natural world. Seeds similar to those I have discerned can likely be found in any craft, occupation, or lifestyle that calls to you and causes your heart to sing. You can find the seeds by following the threads that connect the physical or intangible aspects of your work back to their sources, by directly learning about the diversity of materials, tools, and ingredients that go in, and experiencing with your own senses what comes out. When you find the seeds that help change your worldview, that feed your soul, I encourage you to gather them and consider how they can be kept alive. They won't stay viable for long if locked away, hoarded and hidden from others. The seeds can't be forced to stand still in the darkness, frozen in time, denied the opportunity to change. They survive by adapting, shifting, growing into beautiful plants. Seeds survive by being planted, by being spread, by being passed from one person to the next. Find the seeds of your craft, of your life, and allow them to flourish into heart-rendered expressions of the never-before-heard story of who you are, a unique expression of an animate universe.

At times, the seeds of cheese were given to me without my awareness, dropped in my pocket when I wasn't looking. The keepers of the seeds were happy to give them, and usually did not fully realize the potency of what they were planting in my body when they insisted I come inside and eat at their tables. I have been welcomed into the homes of countless strangers who had less financial wealth than me, yet who fed and housed me without expecting anything in return. I realized that they are in fact the wealthy ones, because they still possess the seeds of true human decency. This hospitality is the default of intact human cultures, and by sitting at these tables, not understanding the words that were spoken, my heart learned another language that changed me, that loosened the stuck cultural grief I was born into. I felt the goodness that all humans are capable of, deep down, below the shields of calcified trauma accumulated by generations of wandering lost and terrified in the dark.

This book is a bundle of seeds that have I gathered and carried around my neck and in my body. These seeds were passed freely to me, and now I bring them back to the land from which I came, to plant them in degraded soils that desperately need true cultivation, a restoration of dignity. The seeds are not simply the microbes, the livestock breeds, tools, techniques, and recipes. They aren't limited to the spaces in which the milk is fermented and cheeses made and aged. The seeds contain our true humanity, our core essence as social primates who tell stories about the stars, give offerings to mountains and rivers, build shrines to feed the forces of nature, and love and argue with our friends and families.

It turns out some of the seeds were actually inside me all along, lying dormant, hidden away for protection from the coldness of a collective insanity. The seeds I've been given awakened those I already carried. By traveling and feeling the wisdom and compassion that humans are capable of, a little water was sprinkled on these neglected seeds, a bit of sun shone in, and sprouts appeared. What grew from those sprouts is my life, which has flowered into this book, a bouquet of words and images crafted by my hands, as an offering to feed the wild holy in you.

I take these seeds I have been gifted,
and drop them into your hands, heart, and mouth.
The seeds do not belong to me or anyone.
They are a part of who we are,
They have been lying dormant in the soil of our hearts,
Waiting for a little opening to allow some sun in,
Waiting for the rain to return.

The old seeds will sprout new orchards and gardens, planted
in the ashes of the past, the ruins of fallen empires.
New ways of living on this planet will sprout with us,
As we reclaim our role as keepers of the seeds.

Now is the time to return to the soil,
to drop the seeds back into the womb of the earth.
To let rain and sun do their work.
So the seeds can sprout, and we can grow again.
As the mycelium of the earth's rind connects us to her, our mother.
Bringing us back together, again.

Acknowledgments

Writing is not hard for me. I don't struggle to find words, which come pouring out like water liberated from a cracked dam. What is difficult is deciding what needs to be left out. Only a fraction of the people I've met and the cheeses I've eaten found their way into this book. My gratitude goes out to all of them.

I thank all the people I have met over the course of my travels who have opened their doors and welcomed me to their tables, whether or not they appear in these pages. My heart goes out to the border crossers, to the cranks who live far from town and prefer the company of four-legged animals, to the folks who do whatever they need to do to keep farming, to care for the land and the animals.

My gratitude goes to everyone who told me I should write a book, who encouraged me to share this story as it unfolded, as well as the team at Chelsea Green for their hard work in allowing my ideas and reflections to manifest in physical form. I would never have walked into this pasture without the prodding of your stick.

And my hat goes off to all the knowledge holders who have freely shared their wisdom, skills, and seeds with me.

May we all dance together someday, as the light of dawn returns to the sacred earth.

Glossary

abomasum, abomasa (plural): The organ from which rennet is made, which is harvested from a milk-drinking ruminant. It is the fourth compartment of a ruminant's digestive tract and secretes enzymes that coagulate milk.

acid-and-heat coagulation: A process seen in cheeses such as ricotta and paneer, involving the denaturing of proteins. It is one of the major rennetless technologies used to coagulate milk in cheesemaking. Milk or whey is heated to near boiling and an acid is added (vinegar, citrus juice, citric acid, or sour whey), resulting in the formation of a delicate curd.

agropastoralism: A sustenance strategy that combines the raising of crops with the keeping of livestock. One type involves settled pastoralists who grow a limited number of fruits and vegetables, and keep small herds that can be fed on or near their farms. Another type involves mobile pastoralists who bring large herds onto fields where annual crops are grown or into orchards and vineyards during the off season, when fields are fallow or orchards not fruiting. The livestock feed off the stubble of grains, vegetable detritus, fallen fruit, or cover crops while depositing fertilizer and working the soil.

alpage: A French term for high-elevation pasture where livestock are brought in the summer. It can also describe a cheese made at an alpage, or the habitations found there. The term is used across the Alps and has synonyms in many languages, such as *seter* in Norwegian and *planina* in Slovenian. It is generally part of a vertical transhumance from lower-elevation settlements.

backslopping: See **carrying forward**.

biofilm: A thin layer of microorganisms and their byproducts that can form on hard surfaces or body parts, such as our teeth. In cheese, we can consider a rind to be a well-developed biofilm, and biofilms often exist on milking and cheesemaking equipment. Biofilms are frequently seen as negative; modern dairy sanitation attempts to prevent them from forming, and eradicates them when they do via chemical sanitation regimes. I point to positive biofilms that have been shown to be benign or beneficial when existing on wood cheese vats, on aging boards, and on various surfaces of barns, milking parlors, and creameries.

bloomy rind: A style of cheese defined by a fluffy, white rind on a small, high-moisture cheese. Famous examples are brie and camembert, small lactic goat cheeses such as crottin and Valençay, or robiolas. Most are marked by pure white growth of tall, downy *Penicillium camemberti*, but some have rinds where *Geotrichum candidum* flourishes. The growth of the rind fungus breaks down the paste inside the cheese, causing it to soften and take on flavors of hay, mushrooms, and damp cellar.

carrying forward: An alternative to the term *backslopping*, which is derogatory and implies a lack of hygiene. Carrying forward is a traditional approach to culturing fermented foods where a small amount of a previous batch that contains thriving populations of desirable microbes is added to a fresh batch to inoculate it. This is seen in sourdough, kombucha, and natural cheese starters such as clabber, kefir, and whey. Wooden cheese vats, forms, and tables can also serve as reservoirs of active culture, carrying these communities forward when cheese is made daily.

caseificio: An Italian word for a creamery, or a place in which milk is processed into cheese.

clabber: A thick sour milk similar to kefir that can be used as a natural cheese starter, home to a culture originating from raw milk. It is maintained by daily feeding, which carries forward the culture. Milk has clabbered when it has coagulated without any added rennet, due to the level of acid produced by the fermentation.

coagulum: The gel resulting when liquid milk coagulates via the action of rennet or acid. It is a homogeneous mass until it is broken or cut into curds and whey. It is sometimes referred to as the curd (as in *cutting the curd*), but *coagulum* is more accurate and causes less confusion, since *curd* can have multiple meanings.

commercial starter: Also known as DVI or DVS, these are blends or single strains of bacteria isolated and propagated in a laboratory, then sold in frozen packages by large companies. Many makers defend these as being more reliable and consistent than natural starters, but I feel this is not true, only a way for commercial starters to be marketed. As with GMO seeds, there are serious and seldom-discussed downsides and weaknesses to using commercial starters.

curds: Small pieces of fresh cheese resulting from the cutting of the coagulum. Curds can be cut large or small, and are often stirred in whey until they are firm enough to be made into a single cheese. Once removed from whey, the curds quickly come together in a form that may be pressed or flipped as they release more whey and knit together as a cohesive wheel or other shape.

deep terroir: A term I use for the occurrence of a tangible, distinct, signature aroma in the cheese of a particular place. I use it when I discern an aroma or flavor in a cheese that can also be found on other things, such as an animal, field, barn, or tool. It is an experience of terroir becoming tangible to the senses. See **terroir**.

DVI (direct vat inoculation): See **commercial starter**.

fermentation-produced chymosin (FPC): Also referred to as 100 percent chymosin, this is the most commonly used rennet globally. It results when a gene from a calf stomach is spliced onto a fungus, creating a genetically modified organism that excretes pure chymosin, the main enzyme in animal rennet. It is the rennet equivalent of the commercial DVI starters, a standardized product purchased from corporations.

gastronationalism: The idea that certain foods belong to countries or nations. Food often becomes an object of national pride, a way people identify their nationality and create a sense of unity. Food also becomes a means of stereotyping and denigrating others—for instance, "In the neighboring country they eat horse, which we would never do." There are often overtones of cultural superiority in framing a discussion in these gastronationalistic terms.

Geotrichum candidum (Geo): A fungus that can display mold properties and is commonly found in environments where milking and cheesemaking take place. It grows on the rinds of many styles of cheese and can take on various appearances, as it has many varieties and exists in community with other microbes. When it is the dominant microbe, it often forms a wrinkly, wavy, brainlike rind.

grazing circuits: In many places, livestock are taken on these walks by a herder in order to obtain their food, as opposed to allowing them to graze on their own in fenced areas, or bringing their feed to the barns or yards where they congregate. Grazing circuits often involve moving through a range of plant communities, so livestock obtain a more diverse diet. It is a short-term movement of a few hours or up to one day, with the animals returning to where they started the circuit, a home farm or camp.

heirloom cultures or heirloom starters: Can be maintained by the cheesemaker through carrying forward. Kefir and clabber are examples of heirloom cheese starters. Yogurt can be an heirloom starter, where a spoonful of one batch is used to culture the next.

inquisition: Beyond the historical connotations of forced conversion to Christianity and repression of animistic folk beliefs and healing modalities, I apply the term to modern efforts to suppress the microbial sovereignty of people who wish to ferment food in their own homes and feed their communities outside the channels sanctioned by the state. The inquisition is a tendency of imperial power structures, national governments, and broad economic forces to view the political and food sovereignty of peasants and pastoralists with a contempt that they act upon with legal, political, or armed violence.

kefir: A sour milk beverage and natural cheese starter culture that is similar to clabber but has physical structures that are home to the SCOBY responsible for fermentation. Called grains, these structures are small clusters of sugar chains that cement together dead microbe cells and have various architectures, often resembling cauliflower or the fractal spiraling of romanesco. The community of microbes creates more of these grains, upon which they live. When the grains are put in fresh milk, a fermentation occurs resulting in souring and coagulation into a thick probiotic beverage.

LAB: Lactic acid bacteria. A broad category of bacteria from many genera that ferment sugars into lactic acid. They are a part of the raw milk microbiota but can vary greatly in their abundance. They are also a part of the fermentations of sourdough, lacto-fermented vegetables, and krauts, and can be found in certain types of beer.

lactic acid: The main byproduct of milk fermentation. Bacteria feed on sugars and release lactic acid, which results in the milk or cheese becoming increasingly sour. This can be registered as sourness on the tongue, or as aromas of sweet and sour fermenting milk that are described as being lactic.

lactic coagulation: There are two types, pure and hybrid. Pure lactic coagulation involves no rennet; milk gets so sour through fermentation that it shifts from a liquid to a gel with a pH of around 4.6, in the case of yogurt, kefir, or clabber. In cheesemaking, hybrid lactic coagulation and the resulting lactic cheeses (chèvre, fromage blanc) utilize a small dose of rennet, leading to a long coagulation of twelve to twenty-four hours. The curd is very sour at the time of cutting. In this sense, lactic coagulated cheeses can be considered a hybrid of pure lactic and rennet coagulation approaches.

lactose: The main sugar in milk. It is a disaccharide, meaning it consists of two sugar molecules bonded together. It's still considered a simple carbohydrate and is broken down in the initial stages of milk's fermentation into glucose and galactose, which are fermented into acids and other compounds, or transformed by enzymes. Some bacteria can digest lactose directly.

lipases: Enzymes that break down fats, which can be produced by starter cultures or found in raw milk. A more significant source is from rennet, and some rennet pastes contain high levels of lipases. These can lead to distinctively sharp, piquant, spicy flavors in aged cheese.

Maillard reactions: A series of chemical reactions between sugars and amino acids from proteins when foods are heated. They occur during the browning of meat in a hot skillet. Sweetness and umami are elevated, and flavors reminiscent of caramel and butterscotch can emerge, although the process differs from true caramelization.

mesophilic: Bacteria can be divided into broad categories by the temperature at which they thrive. Some grow rapidly at low temperatures, other at high. Mesophilic bacteria or mesos prefer a medium range of 68–102°F (20–39°C). Most small, soft cheeses have a fermentation dominated by mesos, as do kefir and clabber.

microbial ecology or microbial ecosystem: The interactions of various microbiota, similar to the interactions of diverse species in a forest. This could include the original milk microbiota as it interacts with the microbiota on milking equipment, on human bodies, airborne in a barn or milking parlor, or residing in a cheese-aging space. For me, the term has connotations of the relationships between microbes, their cooperation or competition, balance, and shifting ratios. A farm or individual cheese can be spoken of as having its own microbial ecosystem.

microbiota: The collection of living microorganisms living in a defined environment, such as oral, anal, gut, or mammary microbiota. The term *milk microbiota* refers to the community of microbes living in milk. I use it mainly to refer to fresh, sweet milk immediately after it leaves the body of the animal, at which point it begins to ferment and is impacted by wider conditions and various other environmental microbiota. The term can be singular or plural.

milk fat globule (MFG): The structure in which fat exists when diffused in liquid milk. MFGs are broken apart by homogenization, which prevents fat from rising as cream. They can also be ruptured in the butter-churning process, causing cream to separate into butter and buttermilk.

milled curd: A family of cheeses made by breaking up blocks of freshly made cheese, mixing salt into the pieces, and pressing them together into a final shape. The United Kingdom is home to many famous milled-curd cheeses that are now made globally, such as cheddar. Various implements are used to break curd into different sizes and textures. This can be done by hand, knife, meat grinder, apple mill, or purpose-built machines such as a peg mill.

naturally fermented: In the context of this book, a cheese or milk beverage made without commercial starter cultures. This includes cheese made with clabber, kefir, whey, or in a wooden vessel that harbors beneficial LAB. It also includes cheeses made without an added starter culture.

natural rinds: Rinds on cheeses that develop with minimal intervention, with a range of yeasts, molds, and bacteria taking up residence. Natural rinds are often multicolored and rustic in appearance.

pellicle: A visible biofilm that can form on the surface of many fermented liquids. The most famous example is the cellulose layer that forms on top of kombucha. Many refer to this as the mother or SCOBY, but the pellicle is a barrier that is made by the community, and a home for the SCOBY, which lives in the liquid and congregates in and on the pellicle. Like kefir grains, the pellicle is similar to a coral reef in that it is built by members of a biotic community and provides habitat for this community. The reef, or pellicle, is not the community per se, but where it lives. See **SCOBY**.

rennet curd: The basic process for most cheeses is to coagulate milk that is still sweet with a high dose of rennet, forming a strong coagulum quickly (in thirty to sixty minutes). This is opposed to the older techniques of acid and heat, or pure lactic coagulation.

SCOBY: The acronym for "symbiotic community of bacteria and yeast." Used frequently in the production of kombucha, SCOBYs can also be found in sourdough and many fermented foods. Kefir and clabber are SCOBYs. These communities are generally microscopic but can sometimes become visible as a biofilm or pellicle, which serves as a vehicle and refuge for the community.

scotta: The whey remaining after ricotta is made. Also called second or de-albuminated whey. It is usually clear and green, and is often used to feed hogs or other livestock.

smear rind or smear ripened: A cheese that undergoes ripening marked by a community of bacteria that form a sticky, orange-red pigmented rind that is highly aromatic. Sometimes called washed-rind cheese or stinky cheese. The smear is usually formed by washing with a salt brine, as the communities that make up the smear are often composed mainly of salt-tolerant bacteria.

spontaneous fermentation: A process where something is allowed to ferment without the addition of a starter culture. Fresh raw milk can be left to spontaneously ferment, which is how clabber is initiated. Cheeses can be made without adding a starter, relying on spontaneous fermentation. This is also referred to as a wild fermentation.

sweet milk: Milk that has its sugars intact, prior to the onset of fermentation. This is what is found in the grocery store and what many people consume as drinking milk. It requires lots of energy to keep milk in this sweet state, to prevent it from fermenting into healthier and more shelf-stable sour-milk beverages or cheese. I use the term to distinguish it from sour milk, and to point out that the consumption of sweet milk is a modern phenomenon.

terroir: This French term is popularly used in winemaking to refer to the combination of factors that lead to a wine having a distinct flavor profile specific to a place. The factors include climate, soil, and exposure to sunlight. *Terroir* is used more broadly to refer to a taste of place, which can manifest in many foods. I identify terroir in cheese on many levels, such as botanical, microbial, and cultural.

thermophilic bacteria: Bacteria that grow rapidly at high temperatures of 104–113°F (40–45°C), with many able to survive above this range. Thermophilic bacteria or thermos are cultivated as the main drivers of fermentation in many yogurts, and in large wheels of alpine and Grana-style grating cheeses such as parmesan and Grana Padano.

tomme: A vaguely defined term referring to a small- to medium-sized wheel that is somewhere between soft and hard, often with a natural rind colored by molds.

transhumance: The seasonal movements of humans and livestock across landscapes. The most common form is a vertical transhumance, in which animals from lower-elevation permanent settlements are brought up to mountain pastures in the summer. There are also horizontal transhumances, where movements stay at similar elevations but go from north to south, or between different climate zones.

whey: The protein-rich liquid that remains after cheese is made.

Further Reading

Asher, David. *The Art of Natural Cheesemaking.* Chelsea Green Publishing. 2015.

Asher, David. *Milk Into Cheese.* Chelsea Green Publishing. 2024.

Blaser, Martin J. *Missing Microbes.* Picador. 2015.

Caldwell, Gianaclis. *The Small-Scale Dairy.* Chelsea Green Publishing. 2014.

Corbett, Jim. *Goatwalking.* Viking. 1991.

Dunn, Rob. *Never Home Alone.* Basic Books. 2018.

Jackson, Robyn. *Cheese from Scratch.* Homestead Living. 2025.

Katz, Sandor. *Sandor Katz's Fermentation Journeys.* Chelsea Green Publishing. 2021.

Kessler, Brad. *Goat Song.* Simon and Schuster. 2009.

Khosrova, Elaine. *Butter: A Rich History.* Algonquin Books. 2016.

Kindstedt, Paul. *Cheese and Culture.* Chelsea Green Publishing. 2012.

Lee, Victoria. *Arts of the Microbial World.* University of Chicago Press. 2021.

Luniova, Katya. *The Art of Clabber.* Sawdust Publishing. 2025.

Mendelson, Anne. *Milk.* Alfred A. Knopf. 2008.

Mendelson, Anne. *Spoiled.* Columbia University Press. 2023.

Palmer, Ned. *A Cheesemonger's History of the British History of the British Isles.* Profile Books. 2019.

Percival, Francis, and Bronwen Percival. *Reinventing the Wheel.* University of California Press. 2017.

Provenza, Fred. *Nourishment.* Chelsea Green Publishing. 2018.

Saladino, Dan. *Eating to Extinction.* Farrar, Straus, and Giroux. 2022.

Sarangerel. *Riding Windhorses.* Destiny Books. 2000.

Yong, Ed. *I Contain Multitudes.* Ecco/HarperCollins. 2016.

Index

M

N

About the Author

Trevor Warmedahl is a cheesemaker, fermentation educator, author, and photographer who writes about the relationships among milk microbes, livestock, landscapes, and human cultures. His project Milk Trekker involves documenting traditional dairying and cheesemaking practices in various countries. Sour Milk School is his mobile educational endeavor to spread information about natural cheesemaking and the philosophy of reclaiming our role as stewards of microbes. An avid ultralight backpacker and world traveler, Trevor dedicates his time to visiting cheesemakers, dairy pastoralists, chefs, farmers, and fermenters, learning while volunteering for them, and sharing his knowledge during cheesemaker residencies and educational collaborations. He posts on Instagram and Substack as Milk Trekker.